Particle Accelerators: A Brief History

M. Stanley Livingston

Associate Director, National Accelerator Laboratory

Harvard University Press : Cambridge, Massachusetts : 1969

Distributed in Great Britain by Oxford University Press, London
Library of Congress Catalog Card Number 69-18038
Printed in the United States of America

Preface

Particle accelerators are among the most useful tools for research in nuclear physics and in high-energy particle physics. The rapid growth of these research fields has been due, in large measure, to the development of a sequence of electronuclear machines for acceleration of ions and electrons. The high-intensity and well-controlled beams from these machines can be used to disintegrate nuclei, produce new unstable isotopes, and investigate the properties of the nuclear force. Modern high-energy accelerators can produce excited states of the elementary particles of matter, forming new unstable particles with mass values much higher than those of the stable particles. Fundamental questions can be asked of nature, and answered by experiments with these very high-energy particles. The field of high-energy particle physics is on the threshold of a significant breakthrough in our understanding of the particles of nature and the origins of the nuclear force.

Energies achieved with accelerators have increased at an almost exponential rate during the past 35 years. The field has been characterized by a sequence of new concepts or inventions, each leading to a new machine capable of still higher energy, and each stimulating the development and construction of a new generation of accelerators. At times, the new developments came so fast that it was difficult to determine which laboratory or machine held the current energy record. A chronology of the major new steps in the development is included as an appendix, which lists the new concepts, first operations of new types of accelerators, and new energy records. Also included in the Appendix is a graph of the growth of energies achieved by accelerators during the years 1930 to 1968, with extrapolations into the future.

This monograph is largely derived from manuscripts of talks and lectures delivered at Harvard University. Four chapters were presented as Morris Loeb Lectures at the Harvard Physics Department from April 8 to 18, 1968. One chapter is a talk given at the Physics Department Colloquium on December 11, 1967. And the final chapter is based on a National Accelerator Laboratory Report, NAL-12-0100, dated June 18, 1968.

These chapters represent facets of accelerator history as experienced and observed by the author. The monograph is not intended as a complete history of all accelerator development, and several significant types are only briefly mentioned, although they have played major roles. Rather, this is a limited compilation of those phases of accelerator development with which I have had close associations. It is to be hoped that experts in other accelerator categories will also tell their stories of the origins and development of these machines.

For a more detailed description of the technical development of accelerators the reader is referred to M. S. Livingston and J. P. Blewett, *Particle Accelerators* (McGraw-Hill, New York, 1962). A more complete report on the history of accelerator development can be found in M. S. Livingston, Ed., *The Development of High-energy Accelerators* (Classics of Science Series, G. Holton, Gen. Ed.; Dover, New York, 1966).

M. Stanley Livingston

Oak Brook, Illinois
June 1, 1968

Contents

[1]

The Race for High Voltage

When J. D. Cockcroft and E. T. S. Walton first disintegrated lithium nuclei with accelerated protons of 500 kilovolts energy, at the Cavendish Laboratory of Cambridge University in 1932, they opened a new era in science. This event may well be considered the origin of modern nuclear physics. It can also be taken as the starting point in accelerator history.

A decade earlier in the same Laboratory, Ernest Rutherford had disintegrated nitrogen nuclei for the first time, using the naturally occurring alpha particles from radioactive elements which had energies of about 5 million electron volts (MeV). During this decade much had been learned to confirm the nuclear character of atoms and to determine the magnitudes of nuclear binding energies. In a speech before the Royal Society in 1927 Rutherford [1] expressed his hope that charged particles would some day be accelerated to energies exceeding those of radioactive radiations, capable of disintegrating nuclei. Scientists in several other laboratories held similar views of the need for high-speed particles and of their value as probes for studying atomic nuclei. Despite the fact that the goal of several million volts seemed well out of reach of known techniques for producing high voltages, by 1929 work had started in several laboratories in several countries to develop the electrical machines needed. This competition soon took on the aspects of a race for high voltage, and for the devices needed to accelerate particles to energies sufficient to disintegrate nuclei. The initial goal was one million volts (1MV).

HIGH-VOLTAGE MACHINES

During the next few years many of the known techniques for producing high voltage were studied to see if they could be extended to even higher potentials. "Surge generators" had been developed for testing electrical equipment at very high voltages. These consisted of stacks of capacitors which were charged in parallel from a dc potential supply and then discharged through cross-connected spark gaps to develop surges of high voltage of a few microseconds duration. The highest-voltage surge generator was built in the General Electric Co.[2] plant at Pittsfield, Mass., in about 1932; it was capable of producing voltage surges of over 6 MV, and was used for breakdown tests of electrical insulators and other equipment.

In 1930 A. Brasch and F. Lange [3] in Germany applied 2.4-MV voltage pulses from such a surge generator to a crude vacuum chamber made of alternate rings of metal and fiber tightly compressed between end plates. The peak surge current was of the order of 1000 amperes; the discharge tube practically exploded on each surge and had to be cleaned and reassembled frequently. A metal-foil window at the grounded end allowed a beam of high-energy electrons and gas ions to emerge into the air, where it produced an intense blue glow extending outward as far as 1 meter. Presumably this discharge was attended by some nuclear disintegrations, but they were not identified.

An even more extreme approach was an attempt to utilize the high potentials developed in the atmosphere during electrical storms. In 1932 C. Urban and others stretched an insulated cable across a valley between two peaks in the Alps. From this cable a spherical terminal was suspended. During thunderstorms high potentials would develop between this terminal and the valley floor; sparks several hundred feet long were observed. Plans had been made to install a discharge tube for the acceleration of particles but were abandoned when Dr. Urban was killed by lightning.

Another engineering test installation was developed by R. W. Sorensen [4] at the California Institute of Technology in the early 1920's, for the Southern California Edison Company, which was using 220-kilovolt lines at 50-cycle frequency for long-distance transmission of electric power. In this technique three 250-kV transformers were arranged in series and mounted on insulating platforms; an "exciter" winding at the high-potential end of the secondary of one transformer was used to supply the primary winding of the next transformer. With this arrangement the rms potential of the high-voltage terminal of the third transformer was 750 kV above ground and the peak voltage exceeded 1 MV. The installation was used for years for the study of high-voltage breakdown of electrical devices. In about 1928 this system was taken over by C. C. Lauritsen [5] and his associates of the California Institute of Technology, to be used as a voltage source for acceleration of particles. They first developed X-ray tubes operating at potentials up to 750 kV, using a single large porcelain insulator for the vacuum chamber. By 1934 they developed a positive-ion accelerating tube with potentials up to 1 MV and started a program of nuclear research. Eventually, the cascade transformer was replaced by a belt-charged electrostatic generator, which was found to be more suitable for nuclear experiments.

The Tesla coil is a resonance transformer which can generate oscillatory pulses of high potential. In its usual form the primary circuit has large capacitance and low inductance, and is excited by the discharge of a series spark gap; the secondary is wound of many turns with high inductance and low distributed capacitance, with the same resonant frequency as the primary. In the early 1930's a group at the Department of Terrestial Magnetism of the Carnegie Institution of Washington, consisting of M. A. Tuve, G. Breit, O. Dahl, and L. R. Hafstad,[6] attempted to develop the Tesla coil as a voltage source for positive-ion acceleration, with the secondary coil immersed in oil for insulation. They reported peak potentials of up to 3 MV. They also developed discharge tubes in which the applied potential was divided between multiple tubular electrodes; this became an accepted technique in the future. However, the oscillatory character of the potential obtained from the Tesla coil made it unsuitable for particle acceleration. The Carnegie group abandoned it in 1932 in favor of the belt-charged electrostatic generator developed by Van de Graaff.

Several modifications of the resonance transformer were developed by others. In 1933 D. H. Sloan [7] at the University of California built a radiofrequency resonance device operating at 6 megacycles per second or 6 megahertz (MHz). It was used as an electron accelerator to generate X-rays of up to 1.25 MV; an installation of this type built by M. S. Livingston and M. Chaffee gave many years of service at the University of California Hospital in San Francisco. Another type, resonant at 60 Hz, was developed by E. E. Charlton [8] and associates at the Schenectady laboratory of the General Electric Company in 1934, which was also used as an X-ray generator at potentials up to 1 MV.

All the devices described above have the limitation of producing either pulsed or alternating potentials, and this is their basic fault as ion accelerators. Several of them have been reasonably successful as electron accelerators for the production of X-rays, where the stability requirements are not so severe and the X-ray tube acts as a rectifier; but they have all failed as sources of positive ions.

The successful techniques that have survived in the competition are those which develop steady direct voltages and which can be regulated to maintain constant voltage with good precision. A system that has succeeded in this respect is the common direct-current power supply, consisting of an alternating-current transformer, a rectifier, and a filter circuit to smooth out the voltage ripple. Although the transformer-rectifier circuit had been available for many years, the voltages to which it had been developed (less than 100 kV) were far too low to be useful in nuclear physics. However, a variation known as the voltage-doubling or voltage-multiplying circuit can produce considerably higher potentials; it had been developed [9] and used in the early 1920's for application to high-voltage X-ray tubes.

THE VOLTAGE MULTIPLIER AND
THE FIRST DISINTEGRATION

Cockcroft and Walton of the Cavendish Laboratory were inspired by Rutherford to search for a device of modest size and energy which might still be sufficient to disintegrate nuclei, and chose to develop the voltage-multiplier circuit to higher voltages; they planned their system to operate at 700 kV or higher. Then in 1929 G. Gamow,[10] and also E. U. Condon and R. W. Gurney,[11] using the new theoretical tools of wave mechanics, showed that protons of relatively low energy (as low as 500 keV) should have a reasonable probability of penetrating the potential barriers of light nuclei. This more modest goal seemed feasible and justified Cockcroft and Walton's choice. Accordingly, they initiated experimental studies of nuclear disintegration when they had achieved only 500 kV. They used the lightest practicable target (metallic lithium) and initially employed a scintillation-counting technique, similar to that used by Rutherford in his early alpha-particle experiments, to observe the charged fragments of disintegration (He nuclei). They observed the disintegration of Li by these 500-keV protons early in 1932. Their results were reported in a series of papers in the *Proceedings of the Royal Society*,[12] which gave full details of the technical development, voltage calibrations, and experimental observations. These papers are classics in nuclear physics and have brought enduring fame to the authors, who were awarded the Nobel Prize in physics for 1951.

The Cockcroft-Walton voltage-multiplier circuit (Fig. 1) is an alternating-current circuit which uses rectifiers to charge capacitors in parallel at low potential, and discharges them in series through a load resistor to develop the high potential. The arrangement at the Cavendish used vertical stacks of capacitors and rectifier tubes, with rounded corona shields at the high-voltage terminals (Fig. 2). Cockcroft and Walton obtained a fourfold voltage multiplication, with a steady dc output of about 500 kV. The ion-accelerating tube was also a vertical array of tubular electrodes within a vacuum chamber

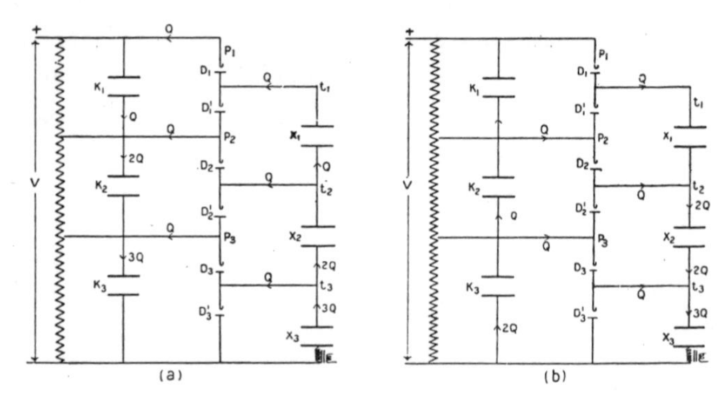

FIG. 1. Schematic circuit diagram of the voltage multiplier, showing the direction of charge flow during the rectification cycle. The output of a single primary transformer (lower right) provides the charge, which is distributed between the capacitor stacks in successive cycles.

formed of insulating glass cylinders. The ion source at the upper terminal consisted of a high-voltage discharge in hydrogen gas in a separate chamber with a small-diameter exit hole through which the hydrogen ions emerged into the accelerating tube. After acceleration down the column, the beam of hydrogen ions emerged through a channel in the grounded baseplate, where it was directed against the target (Fig. 3).

The voltage multiplier has been widely copied and modified in other laboratories, but seldom for potentials higher than 1 MV. The Philips Company of Eindhoven, Holland, built a well-engineered system for the Cavendish Laboratory which operated at 1.25 MV for many years. But the primary use in recent years has been as a preaccelerator of particle beams for injection into higher-energy machines.

THE ELECTROSTATIC GENERATOR

A major contender in the race for high voltage in the early years was the electrostatic generator. R. J. Van de Graaff started the development of the belt-charged generator at

FIG. 2. Voltage multiplier with which Cockcroft and Walton achieved nuclear disintegration in 1932.

Princeton in 1930, after becoming interested in the need for high-voltage machines as a Rhodes Scholar at Oxford in 1928. In its modern pressure-insulated form it has become so successful as a voltage source that it has displaced most other types of direct-voltage generator. In the energy range of 1 to 9 MV to which it has been developed it can deliver a steady, parallel beam of particles with excellent energy uniformity and stability. It is by far the most popular type of accelerator and the number of installations around the world is conservatively estimated to be over 500.

Van de Graaff [13] described his first model of a belt-charged

FIG. 3. J. D. Cockcroft in the basement room below the voltage multiplier where nuclear disintegrations were observed.

generator before the American Physical Society in 1931. Two spherical aluminum terminals, 24 in. in diameter, were mounted on 7-ft glass rods, each with a motor-driven silk belt to transport the charge. The belts were charged with ions produced by corona discharge from needle points, and discharged with similar points in the terminals. One sphere was given a positive charge and the other a negative charge. A potential difference estimated to be 1.5 MV, limited only by sparking or corona discharge, was developed between terminals. The simplicity of the principle, and the fact that a steady direct voltage was produced, made this technique attractive; the possibility of extending it to higher voltages was evident. Groups in several laboratories became interested and joined in the development.

In 1932 Van de Graaff went to the Massachusetts Institute of Technology where he started designs for a really large generator.[14] It consisted of two 15-ft spherical terminals mounted on vertical insulating fiber cylinders each enclosing a 4-ft-wide belt for charging, one carrying positive and the other negative charges. A discharge tube for accelerating ions was to be supported between the terminals, and the laboratory for observing experiments was to be within one of the 15-ft shells. The concept of the project was exciting and the scale tremendous for that time. When operated, sparks up to 25 ft long would occasionally jump between terminals or down the columns. When the terminals were charged, the static fields would literally raise the hair of a person standing at the base. I remember vividly the experience of standing within one sphere while both were being charged; monstrous sparks would jump the gap and strike the rim of the porthole just a few inches away. However, unclean conditions in the aircraft hangar near the ocean where the generator was first assembled, and difficulties in mounting the horizontal discharge tube, were extreme, so that the generator never performed satisfactorily as an accelerator. In 1937 it was moved to MIT and located in an enclosed building with the two terminals in contact (Fig. 4). One column was used for the charging belt and the other for a vertical accelerating tube so experiments could be performed in a room beneath the floor. By 1940 the modified accelerator was in operation at 2.75 MV,[15] and it gave many years of valuable service as a research tool. It eventually became obsolete and was moved to the Boston Museum of Science as a permanent exhibit.

Meanwhile, the group headed by M. A. Tuve[16] at the Carnegie Institution of Washington adopted the electrostatic generator as a voltage source. With the advice and cooperation of Van de Graaff they built first a 1-meter sphere mounted on a

FIG. 4. C. L. Van Atta, D. L. Northrup, and L. C. Van Atta beneath the 15-ft terminals of the 2.7-MeV electrostatic generator at the Massachusetts Institute of Technology. (From *Particle Accelerators* by M. S. Livingston and J. P. Blewett. Copyright 1962 by McGraw-Hill Book Company. Used by permission of McGraw-Hill Book Company.)

tripod of insulating struts, with a vertical accelerator tube formed of glass tubing. By 1933 (only a year after Cockcroft and Walton had observed disintegrations) this first practical electrostatic generator was in operation producing hydrogen ions of 0.6-MeV energy, and was soon put to use for nuclear experimentation. A 2-m generator was built next, with a 1-m inner shell as a voltage divider, which was in operation at 1.3 MeV by 1935 (Fig. 5). Major contributions of the Carnegie group were their precise voltage calibrations and studies of the proton range-vs.-energy relation. They used magnetic deflection of the emergent beam of protons, and current measurements with a column of calibrated resistors, to determine proton energy. They measured and published values of many sharply defined nuclear resonances,[17] which became substandards for cross-calibration of the results from different laboratories.

Meanwhile, other groups exploited the use of high gas pressure for insulation of the terminal, by enclosing the generator in a pressure housing. With this technique much higher voltages were ultimately obtained. A group at the University of Wisconsin led by R. G. Herb pioneered the development with a series of pressure-insulated generators of steadily increasing energy, which supported a continuing program of research in nuclear physics. The Herb design used concentric terminal shields to distribute the voltage, mounting the generator from a grounded baseplate, and with a horizontal discharge tube; the pressure housing rolled back on rails to open the chamber for servicing. The series started with a first model which reached 0.75 MV in 1935,[18] a second operated at 2.4 MV by 1938,[19] and a third reached 4.0 MV by 1940.[20]

Another development was the vertically mounted pressure-insulated generator which was started at the Massachusetts Institute of Technology by J. G. Trump[21] of the Electrical Engineering Department, Van de Graaff, and their associates. This group made a comprehensive study of the engineering problems of high dc potentials, and developed a sequence of

FIG. 5. View of the 2-m electrostatic generator and provisional discharge tube at the Bureau of Terrestial Magnetism. Shown are O. Dahl (on ladder), a technical helper, L. R. Hafstad and M. A. Tuve.

generators for research use at MIT. This development culminated in the present 9-MV machine, which has reached the highest potential of any single-stage generator.

In 1947, Trump, Van de Graaff, and others formed the High Voltage Engineering Corporation, for the commercial production of "Van de Graaff" generators. This company has produced a series of models of pressure-insulated machines of increasing energy and reliability, and is the accepted leader in the technical development of electrostatic generators. An advanced development of this company is the "tandem," a double-ended, horizontal generator in which negative hydrogen ions are accelerated "up" to the central terminal where they are stripped of electrons in a gas jet to become positive ions and are then accelerated "down" to ground potential to obtain particle energies of double the terminal voltage.

THE CYCLOTRON

A close second in the race to achieve nuclear disintegration was Ernest O. Lawrence, who with his students at the University of California built the first cyclotron. The significance of this development was in the use of resonance acceleration, in which high-speed particles were produced without the need for high voltages. This new principle avoided many of the technical limitations encountered in the use of direct high voltages due to breakdown of insulation, and opened the way to a new category of accelerators capable of very much higher particle energies.

Lawrence conceived the idea on reading a paper by R. Wideröe,[22] who described an experiment in which charged particles traversed tubular electrodes, in resonance with a radiofrequency electric field applied to the electrodes, and emerged with energies twice what the applied voltage would have given them (see Chapter 2). Lawrence speculated on variations of this resonance principle, including the use of a magnetic field to deflect the particles in circular paths so they would return to the first electrode, where they could utilize the radiofrequency field in many successive traversals; the device would also be more compact in physical size. He discovered that the equations of motion predicted a constant period of revolution, so the particles would be accelerated on each traversal of the gap between electrodes if their motion was in resonance with the applied radiofrequency field. The concept of magnetic resonance was first published in a brief note by Lawrence and N. E. Edlefsen[23] in 1930, without experimental confirmation.

I was a graduate student at the University of California at this time, and this problem was suggested by Professor Lawrence as the subject for a research investigation to demonstrate the validity of the resonance principle. The study was successful and resulted in a doctoral thesis[24] dated April 14, 1931, which was not published. This first model used small-sized laboratory equipment including a magnet with 4-in. poles. A flat brass chamber (sealed with wax) which fitted between the poles of the magnet had a single hollow electrode shaped like a capital D to which the radiofrequency potential was applied. Hydrogen ions were formed at the center by flooding the chamber with gas at very low pressure and ionizing it with electrons from a cathode. Resonant ions that reached the edge of the chamber after making scores of revolutions of increasing radius were observed in a shielded collector cup; their energies were confirmed by deflection in a transverse electric field. Cyclotron resonance was observed for hydrogen ions (H_2^+) over a wide range of radiofrequencies and magnetic fields. The most energetic particles obtained were hydrogen ions of 80,000 eV, when the radiofrequency potential applied across the accelerating gap was only 1000 V.

Lawrence moved rapidly to extend this first success to a machine capable of producing nuclear disintegrations. In early 1931 he was awarded a grant by the National Research Council (for $1000) to build a larger machine. And he arranged that

FIG. 6. The first operating cyclotron built by E. O. Lawrence and M. S. Livingston at the University of California in 1931–32. It produced 1.2-MeV protons.

I obtain an appointment as instructor at the University to continue the development. During the summer and fall of 1931, with the guidance of Professor Lawrence, I designed and built a magnet with poles 10 in. in diameter and a cyclotron chamber to fit between its poles (Figs. 6 and 7). By the spring of 1932 this first practical cyclotron was completed and tuned up to accelerate protons to 1.2-MeV energy. The development was described at scientific meetings and was published in a paper by Lawrence and Livingston [25] later in 1932.

FIG. 7. Vacuum chamber for the 1.2-MeV cyclotron, built by E. O. Lawrence and M. S. Livingston, now at the Kensington Museum of Science in London.

A few months before this goal had been reached, the publication by Cockcroft and Walton describing their results became available at Berkeley. Lawrence mounted a concentrated effort to obtain and build instruments that could detect nuclear disintegrations. D. Cooksey and F. N. D. Kurie came from Yale University for the summer to assist in this program. Internal targets were installed inside the chamber of the cyclotron, and a thin-foil window was mounted on the chamber wall outside the target so the particle products of disintegration could emerge into a counter outside the window. Disintegrations in Li and other targets were soon observed; these results were reported in a brief note by Lawrence, Livingston, and M. G. White [26] in September 1932.

Lawrence had already made plans for the next step to still higher energies, and had obtained financial support from the Research Corporation. This next size was the "27-inch." The iron core for the electromagnet was taken from a dismantled Poulsen-arc radio transmitter donated by the Federal Telegraph Company. By the time this machine was completed and operating, in 1933, ions of the newly discovered heavy hydrogen isotope (deuterons) were available, obtained from heavy-water samples provided by Professor G. N. Lewis of the Chemistry Department. This first major Berkeley cyclotron was tuned up (primarily by "shimming" of the magnetic field) to produce 3-MeV and then 5-MeV deuterons, and was described in a publication by Lawrence and Livingston [27] in 1934. An active research program was developed using these particles, exploring the new fields of deuteron reactions, induced radioactivity, and neutron production. A group of young and enthusiastic scientists joined in exploiting this opportunity, including M. C. Henderson, E. M. McMillan, W. Coates, J. J. Livingood, B. B. Kinsey, F. N. D. Kurie, R. L. Thornton, and, in the years to follow, many others.

In 1934 I left Berkeley to join the Physics Department at Cornell University and to build another cyclotron, and a few years later went to MIT to build still another. At Berkeley, Lawrence and his growing team continued the development to higher energies. Their next step was expansion of the magnet poles to 37-in. diameter, which resulted in 8-MeV deuterons by 1936. The potential medical applications of high-energy neutrons, available in significant intensities for the first time, justified and supported the next step, which was the 60-in.

"Crocker" cyclotron; [28] this was completed in 1939, operating first at 16-MeV and later at 20-MeV deuterons or 40-MeV doubly charged helium ions.

The cyclotron took the lead in particle energy in 1932, and held first place for 10 years. Following 1934, cyclotrons were built in many other laboratories, frequently designed by graduates of the Berkeley school. Soon these other laboratories were able to contribute to the development, including the first high-intensity ion source and other features. More than a hundred cyclotrons have been built in laboratories around the world, mostly as sources for nuclear research studies. A few of the larger installations have been supported by medical research funds, and research programs were developed around medical and biological problems such as the use of neutrons for cancer therapy and of radioactive isotopes for tracer experiments in living tissue.

The cyclotron has become a symbol of nuclear physics, and its simple resonance principle is taught in most high school and college physics courses. Small working models have been built by students in high schools. Many of the larger machines were patterned after the 60-in. Crocker cyclotron. The two largest fixed-frequency cyclotrons were the 86-in. at the Oak Ridge National Laboratory and the 225-cm at the Nobel Institute in Stockholm (Fig. 8). Both machines produced protons of 22-MeV and deuterons of 24-MeV energy. Practical limitations restricted the maximum energy to values less than the theoretical limit. This limit to particle energy is due to the relativistic increase in mass of the ions, which ultimately detunes the ions from resonance and prevents further acceleration. However, there is no longer any need to push the cyclotron to this energy limit. Higher energies can be achieved more readily in synchronous accelerators or by use of the special focusing techniques of "isochronous" cyclotrons.

FIG. 8. The 225-cm cyclotron at the Nobel Institute for Physics, Stockholm.

ELECTRON ACCELERATORS

The machine that won the next heat in the race for high-energy particles and took the energy record away from the cyclotron was the betatron, an electron accelerator. The initial motivation for the development of electron accelerators has been somewhat different from that for proton accelerators. In

FIG. 9. The first betatron built by D. W. Kerst at the University of Illinois in 1940. It produced 2.3-MeV electrons.

early years the purpose was to produce high-energy X-rays for medical therapy in hospitals. Such X-ray machines compete with radioactive gamma-ray sources and their output is frequently given in roentgen units. Another application has been found in the metal-products industry for the survey of flaws in castings and welds. Electrostatic generators have been widely used in such applications, in the 1- to 4-MV range.

The first successful betatron was built by D. W. Kerst [29] in 1940 (Fig. 9); it produced 2.3-MeV electrons, and had an X-ray output equivalent to that from 1 gram of radium. Next, Kerst went to the General Electric Company, where with the help of its experienced laboratory staff he built a 20-MeV betatron by 1942. Later he returned to the University of Illinois to build first an 80-MeV "model" and ultimately (1950) a 300-MeV machine,[30] which represents the largest and probably the last large betatron. Before it was completed the electron synchrotron came on the scene; with a lighter and more efficient ring magnet, it displaced the betatron as an electron accelerator in the high-energy range.

In the continuing race for high energy, synchronous accelerators were further developed following World War II. Also, several new types of machine and a major increase in energy were made possible with the introduction of alternating-gradient focusing. For many years the energy record of 33 GeV (10^9 eV) was held by the Alternating-Gradient Synchrotron (AGS) at Brookhaven National Laboratory. This was surpassed by the 76-GeV machine at Serpukov, USSR, which came into operation in time to be announced at the anniversary of the October Revolution in 1967. Plans exist at various levels of authorization for machines of still higher energy. The plan with the highest credibility level at present is for a 200-GeV alternating-gradient proton synchrotron in the United States; a site at Weston, Illinois, has been selected by the Atomic Energy Commission and the staff has been assembled to formulate the final designs for the installation. Other plans exist for 300-GeV and even for 1000-GeV machines. So the race for high energy goes on!

[2]

Ernest Lawrence and the Cyclotron—An Anecdotal Account

I first met Professor Lawrence as a student in his Electricity and Magnetism course when I went to Berkeley as a graduate student in 1929; he was a young associate professor in his second year at the University of California. I was greatly impressed with his enthusiasm and his vivid personality. He seemed always to emphasize the important concepts and conclusions, but took a rather cavalier attitude toward factors of 4π or other details in theoretical developments.

In the early summer of 1930 I asked Professor Lawrence to propose a topic for an experimental thesis. He suggested a study of the resonance of hydrogen ions with a radiofrequency electric field in the presence of a magnetic field—the phenomenon now known as cyclotron resonance. He claimed that the ions should make hundreds of revolutions in the magnetic field, gaining energy in each turn and attaining final energies of 1 MeV or more in a magnet of suitable size. I found that another student, N. E. Edlefsen, who had completed his thesis the previous winter and was awaiting the June degree date, had made a preliminary attempt to observe this resonance phenomenon in the spring of 1930 using a small laboratory magnet, and that Lawrence considered his results very promising. Lawrence described the concept at a meeting of the American Association for the Advancement of Science in Berkeley that spring and submitted a brief article to *Science*.[1]

In discussions with Lawrence in later years I learned that he had conceived the idea of a magnetic resonance accelerator in the early summer of 1929, while browsing through the current journals in the library at the University of California. He saw the illustrations in a paper by Wideröe[2] in the *Archiv für*

Elektrotechnik for 1928, and recognized the resonance principle involved, although he could not read German readily. Wideröe's paper described an experiment in which positive ions of Na and K were accelerated to twice the applied voltage while traversing two gaps at the ends of a tubular electrode to

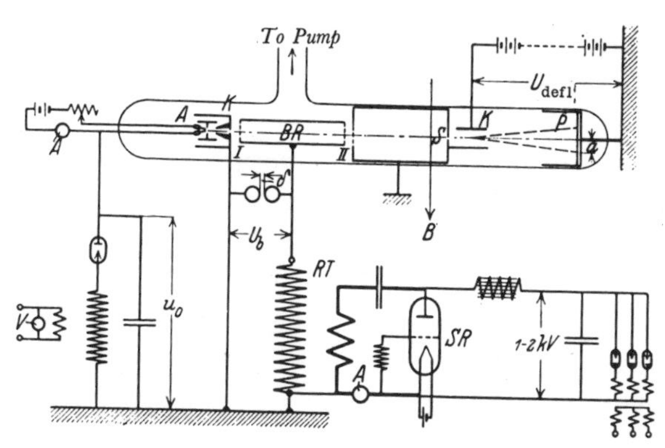

FIG. 10. Diagram of the apparatus used by R. Wideröe in 1928 to demonstrate the doubling of the energy of heavy ions in resonance with a radiofrequency electric field. A radiofrequency generator (lower right) produces rf voltages across two gaps *I* and *II* at the ends of a tubular electrode *BR* through which ions from the source *K* (at the left of *BR*) are accelerated.

which a radiofrequency potential was applied (Fig. 10). This was an elementary linear accelerator. Wideröe described the process as "kinetic voltage transformation." He chose the type of ions, the accelerating frequency, the applied potential, and the gap spacing to achieve resonance. The doubled energy was confirmed by electrostatic deflection measurements on the ions.

Lawrence had been searching for a method of accelerating particles to higher energies than could be attained with dc potentials, in order to study "nuclear excitations." He realized that extension of Wideröe's technique to such high energies would require a very long array of electrodes. He speculated on variations of the resonance principle, including the use of a magnetic field to deflect particles in circular paths so they would return to the first electrode and reuse the electric field in the gap. He found that the equations of motion predicted a constant period of revolution in a uniform magnetic field, regardless of particle energy, so the ions would remain in resonance with an accelerating field of fixed frequency. Charged particles could be made to traverse the same set of electrodes many times, gaining energy on each traversal of the gap between them; the orbit radius would increase as the velocity increased. This was the cyclotron resonance principle and the resonance frequency is now called the cyclotron frequency.

I started experimental work that summer. I first reassembled and recalibrated the 4-in. laboratory magnet used by Edlefsen, built a replacement for the glass vacuum chamber, and studied the broad "resonance" which Edlefsen had observed when the magnetic field was varied. I soon found that this effect was due not to hydrogen ions but probably to heavy ions from the residual gas, which were accelerated once in the radiofrequency field and reached the unshielded detection electrode at the edge of the chamber.

It was now my responsibility to demonstrate true cyclotron resonance. The Physics Department glassblower built for me a sequence of flat glass chambers in which electrodes were mounted on greased-joint seals. Glass was traditionally used for vacuum systems in the laboratory, but this thin, flat glass chamber defied our technical skills. I then built a chamber formed of a brass ring and flat brass cover plates, using red sealing wax for a vacuum seal, in which the several electrodes could be mounted (Fig. 11). The radiofrequency electrode was a single hollow D-shaped half-pillbox facing a slotted bar placed across the diameter of the chamber called a "dummy D." The rf potential was developed by a simple Hartley oscillator; the need for a more efficient rf circuit came later with the effort to increase energy. A 10-W vacuum tube was used as an oscillator and provided up to 1000 V on the electrode, at a frequency that could be varied by changing the number of turns in an external inductance coil formed of copper tubing. Hydrogen molecular ions (H_2^+) were produced through ionization of hydrogen gas in the chamber by electrons emitted from a tungsten-wire cathode near the center of the chamber. Ions that reached the edge of the chamber were observed in a shielded collector cup.

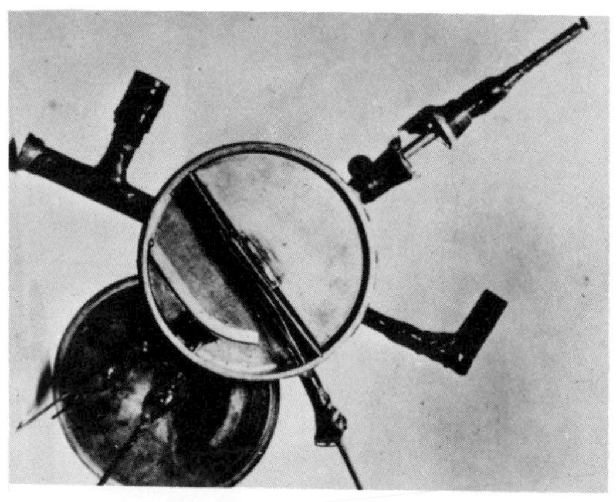

FIG. 11. Vacuum chamber of the first resonance accelerator, used by Livingston to demonstrate cyclotron resonance for a doctoral thesis at the University of California in 1931. Now at the Lawrence Hall of Science in Berkeley.

During the fall I continued the technical development with Lawrence's continued enthusiastic interest and supervision. It was in November 1930 when I first observed sharp peaks in the collector current as the magnetic field was varied over a narrow range. Several techniques were used to prove that the resonance peaks were due to high-energy ions. A deflecting plate and slit system placed in front of the collector gave a rough check of ion energy. But the basic proof was that the magnetic field at resonance was just that calculated from the resonance equation using the measured value of applied radiofrequency. Other resonance peaks observed at lower magnetic fields were explained as due to the 3/2 and 5/2 harmonics of the applied frequency. Incidentally, these resonance values provided a highly accurate check of the magnetic-field calibration. The frequency (measured with a wavemeter) was varied over a wide range and the corresponding values of magnetic field for the resonance peaks were observed. When a graph of wavelength vs. magnetic field was plotted, the points fell on a smooth hyperbolic curve, as predicted by the resonance equation.

The small magnet used for the first studies had a maximum field of 5200 gauss, for which resonance with H_2^+ ions occurred at a wavelength of 76 m; the ion energy was calculated to be 13 keV (kilo electron volts) at the radius of the collector. This goal was reached on January 2, 1931, after working straight through the Christmas and New Years holiday. A stronger magnet was borrowed for a time, capable of producing 13,000 gauss, for which resonance occurred at 30-m wavelength or 10-MHz frequency, and for which the calculated ion energy was 80 keV. This was obtained with an applied peak rf potential of about 1 kV, so the ions traversed a minimum of 40 turns (80 accelerations). This result was reported [3] at a meeting of the American Physical Society early in 1931.

An important consequence of our studies with this prototype was the experimental observation of electric and magnetic focusing. In Lawrence's original conception the electric field inside the hollow rf electrodes should be zero and the field in the gap should be parallel to the plane of the particle orbits. Otherwise it was expected that small transverse fields would produce spiraling orbits that would intersect the electrodes. Accordingly, the electrodes initially had a grid of fine tungsten wires tightly stretched across their apertures at the gap. Resonant peaks were first observed as currents of 10^{-10} and 10^{-11} amperes, requiring our most sensitive electrometers and galvanometers. I knew that these wires were intercepting the circulating beam and felt intuitively that they might not be needed. So while I was on my own during a trip Lawrence made to the East, I removed the grid wires and obtained greatly increased currents, in the 10^{-9}-A range. On his return Lawrence immediately recognized the focusing properties due to the shape of the electric field between open electrode faces (Fig. 12*a*), and we never again used grids. Similarly, when thin shims of iron were inserted in the gap between the chamber and one pole of the magnet, beam intensity was increased for

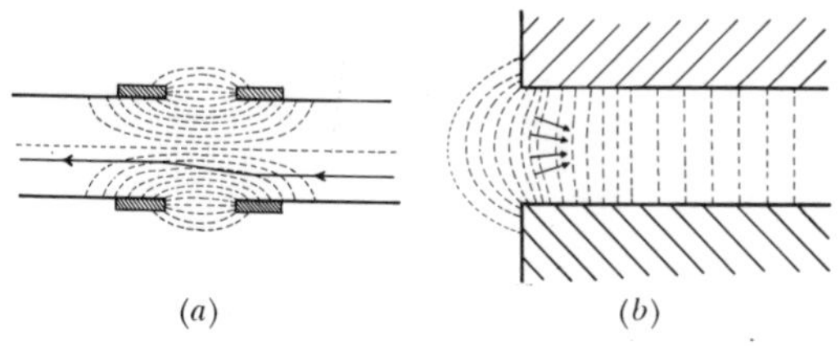

FIG. 12. Sketches used by Lawrence and Livingston in an early publication to illustrate focusing by the electric and magnetic fields (broken lines) in the cyclotron; particles are deflected toward the median plane by both (*a*) the electric field between the D's and (*b*) the magnetic field at the periphery.

certain sizes and locations of the shims. This result caused us to study the effect of the shape of the magnetic field (Fig. 12*b*). The transverse focusing due to the concave-inward shape of the fringing field at the periphery was recognized and checked experimentally by observing the beam thickness at the edge; by means of a probe on a movable greased joint in front of the collector cup this thickness was found to be quite small, of the order of 1 to 2 mm. From then on the "shimming" of the magnetic field became an important and somewhat mysterious technique for tuning up the accelerator.

Lawrence had moved promptly to exploit this breakthrough. In early 1931 he applied for and was awarded a grant from the National Research Council, for $1000, for construction of a machine that could give energies useful for nuclear research. At the end of March, Lawrence told me to start writing my thesis, so I could get my degree, be eligible for an Instructorship in the Department which he had obtained for me for the following year, and could continue the development.

This was just two weeks before the thesis deadline date, but I made it and presented my thesis [4] dated April 14, 1931. Incidentally, I was a poorly prepared candidate. In following Lawrence's enthusiastic lead I had been working nights, weekends, and holidays in the laboratory, with no time for reading or studying. At my oral examination some members of the

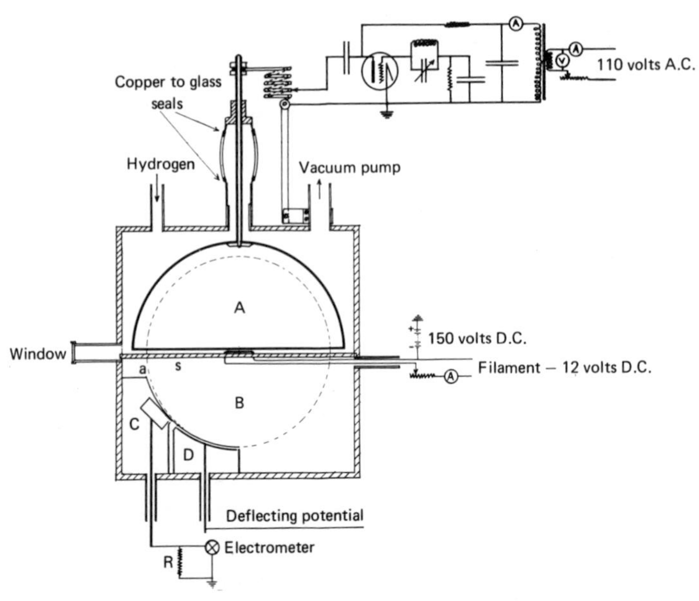

FIG. 13. Diagram of the vacuum chamber for the 1.2-MeV cyclotron built by Lawrence and Livingston at the University of California in 1931. See also Fig. 7.

committee were appalled to find that I had not studied "Rutherford, Chadwick, and Ellis," the basic reference on natural radioactivity, which they considered essential for a person presuming to enter the field of nuclear physics. Again, Lawrence's enthusiasm and personal recommendation prevailed, and I received the degree in May.

During the summer and fall of 1931 I designed and installed a 10-in. magnet and built the other components for a magnetic-resonance accelerator capable of reaching 1-MeV energy, located in Room 339 of LeConte Hall, the Berkeley physics building. As before, the vacuum chamber was a flat brass box and the cover plate was sealed with wax. It contained a single hollow D-shaped electrode for the rf potential, a thermionic cathode, and a shielded deflecting electrode and collector cup located at a radius of 11.5 cm (Fig. 13). The rf oscillator used a 10-kW Federal Telegraph water-cooled power tube in a tuned-plate tuned-grid circuit which produced peak potentials up to 50 kV across the accelerating gap at frequencies up to 20 MHz. I was greatly aided in the development of this first high-power rf oscillator by David Sloan, another graduate student who had been a ham radio operator and was an ingenious student of high-frequency radio techniques. A deflecting electrode was used to draw the beam out of its circular orbit and into the collector; it also provided a rough measurement of ion energy. This first practical cyclotron produced H_2^+ ions of 0.5-MeV energy by December 1931 and H_1^+ ions (protons) of 1.22 MeV, with beam currents of about 10^{-9} A, in January 1932. The progress was reported in several abstracts and a paper [5] was sent to the *Physical Review* on February 20, 1932. This was the first time in scientific history that artificially accelerated ions of this energy had been produced. The original vacuum chamber of this 1.2-MeV cyclotron is now on permanent exhibit in the Kensington Museum of Science in London.

As a personal footnote to history, I recall the day when I had adjusted the oscillator to a new high frequency and, with Lawrence looking over my shoulder, tuned the magnet through resonance. As the galvanometer spot swung across the scale, indicating that protons of 1-MeV energy were reaching the collector, Lawrence literally danced around the room with glee. The news quickly spread through the Berkeley laboratory and we were busy all that day demonstrating million-volt protons to eager viewers.

We had barely confirmed our results and were busy with revisions to increase beam intensity when we received the issue of the *Proceedings of the Royal Society* describing the results of Cockcroft and Walton [6] at the Cavendish Laboratory in disintegrating lithium with protons of only 500-keV energy. At that time we did not have adequate instruments to observe disintegrations. Lawrence sent an emergency call to his friend and former colleague at Yale, Donald Cooksey, who came out to Berkeley for the summer with Franz Kurie. They helped develop the necessary counters and instruments for disintegration measurements. With the help of Milton White, then a graduate student at Berkeley, we installed a target mount inside the collector cup and a thin mica window on the side of the chamber facing the target, outside which counters could be located. Within a few months after hearing the news from the Cavendish we were ready to try for ourselves. Targets of various light elements were inserted, the counters clicked, and we were observing disintegrations. These early Berkeley results [7] confirming Cockcroft and Walton and including several additional targets were published that fall.

Lawrence was planning his next step even before I had completed the 10-in. machine as a working accelerator. His aims were ambitious, but supporting funds were difficult to obtain. He was forced to use many economies and substitutes to reach his goals. In late 1931 he located a magnet core from an obsolete Poulsen-arc magnet with a 45-in. core, and had it machined to form flat pole faces initially tapered to a 27½-in. diameter. The magnet coils were formed of strip copper wound in layers and immersed in oil tanks for cooling. (The oil tanks leaked! We all wore paper hats when working between coils to keep the oil out of our hair.) This magnet was installed in December 1931 in a frame warehouse near the Physics building, later known as the "old radiation laboratory," which was the center of cyclotron activities for many years (Fig. 14). Early in 1932 I turned the 10-in. machine over to White to use for his thesis problem, and applied most of my time to construction of the larger machine.

FIG. 14. E. O. Lawrence and M. S. Livingston beside the magnet frame of the 27½-in. cyclotron while under construction in 1932.

The vacuum chamber was a 27-in. brass ring fitted with iron disks for top and bottom plates; the top "lid" was removable and vacuum-sealed with soft wax. Initially a single rf D was installed, supported by a Pyrex glass insulator, with a slotted bar across the diameter for a "dummy D" (Fig. 15). This al-

FIG. 15. Vacuum chamber of the 27½-in. cyclotron with lid removed, showing the single D electrode and the collector cup set for 10-in. radius.

lowed us to locate the deflection electrode and collector at any chosen radius. The accelerated beam was first observed at small radius, and shimming and other adjustments were made to maximize intensity. Then the collector was moved to a larger radius, and the tuning and shimming were repeated. If we made too large a step and could not find the beam, we made a strategic retreat back to smaller radius. Thus we learned, the hard way, of the necessity for a radially decreasing field to maintain focusing, and produced it with thin disk shims of selected radii placed in the shimming gaps. Eventually we reached a practical maximum radius of 10 in. and installed two symmetrical D's with which higher energies and intensities could be obtained. Technical improvements were added day by day as we gained experience. The progress during this period of development from 1-MeV protons to 5-MeV deuterons was reported [8] in several publications from 1932 to 1934.

I am indebted to E. M. McMillan for a brief chronological account of these early developments on the 27½-in. cyclotron; it seems that earlier laboratory notebooks were lost:

June 13, 1932	16-cm radius, 28-m wavelength, 1.24-MeV H_2^+ ions
August 20, 1932	18-cm, 29-m, 1.58-MeV H_2^+ ions
August 24, 1932	Sylphon bellows installed on filament stem for adjustment
September 28, 1932	25.4-cm, 25.8-m, 2.6-MeV H_2^+ ions
October 20, 1932	Radius fixed at 10 in., two D's installed
November 16, 1932	4.8-MeV H_2^+ ions; beam current 0.001 μA
December 2–5, 1932	Target chamber installed for studies of disintegrations with Geiger counter; start of long series of experiments
March 20, 1933	5-MeV H_2^+, 1.5-MeV He^+, 2-MeV $(HD)^+$ ions; deuterium ions accelerated for first time
September 27, 1933	Neutrons observed from targets bombarded with D^+ ions
December 3, 1933	Automatic magnet-current control installed
February 24, 1934	3-MeV D^+ ions; beam current 0.1 μA; radioactivity induced in C by D^+ bombardment
March 16, 1934	1.6-MeV H^+ ions (protons); beam current 0.8 μA
April–May, 1934	5.0-MeV D^+ ions; beam current 0.3 μA.

These were busy and exciting times. Other young scientists joined Lawrence's group; some worked on accelerator development and others on detection instruments. We joined in teams for taking data and publishing results. In my own list of publications I find 17 abstracts or articles on disintegration results, in addition to several technical papers on accelerator development, during the years 1933–1934. David Sloan and Wesley Coates developed a linear accelerator, using tubular electrodes of increasing length in line, which produced 2.8-MeV Hg^+ ions. Sloan and B. B. Kinsey built a linac for Li^+ ions for energies up to 1.0 MeV. Sloan and J. J. Livingood built a resonance transformer which produced electrons and X-rays of 1-MeV energy. Malcolm Henderson came in 1933; he developed counting equipment and magnet-control circuits, and also spent long hours helping to repair vacuum leaks and on other developments. Incidentally, Henderson invented the name "cyclotron," which was first used only as laboratory slang but was eventually picked up by newspaper reporters and popularized. McMillan joined the group in 1934, and made major contributions to the planning and interpretation of research experiments; he overlapped my time at Berkeley by several months, and is one of only two men still on the Lawrence Radiation Laboratory staff who were there before I left, the other being R. L. Thornton. And we all had a fond regard for Commander Telesio Lucci, retired from the Italian Navy, who was our self-appointed laboratory assistant and a friend to all. As the research results became more interesting we depended heavily on Robert Oppenheimer for discussions and theoretical interpretations. But always Ernest Lawrence was the leader and the central figure, enthusiastic over each new result, intent on each new technical problem, in and out of the laboratory at all hours up to midnight, convinced that we were making history and full of confidence for the years ahead.

One of the exciting periods was our first use of deuterons in the cyclotron. Professor G. N. Lewis of the Chemistry Department had succeeded in concentrating heavy water with about 20 percent of deuterium from battery-acid residues; we electrolyzed it to obtain gas for our ion source. Soon after we had tuned in the first beam of deuterium ions we observed alpha particles from a Li target with longer range and higher energy than any observed in natural radioactivities, coming from the d(Li^6, α) reaction. This led to an intensive program of research on deuteron reactions. We installed a wheel of targets on a greased joint, with targets of many light elements which could be turned into the beam inside the collector cup. A 2-in. re-entrant tube on the side of the chamber, with an end face made of a thin mica foil supported by a grid, allowed us to put our counters close to the target and to measure the range of the charged-particle products.

Chadwick had reported the discovery of the neutron in 1932, and we were aware of its basic properties. As soon as we had developed linear amplifiers capable of observing single particles in thin ionization chambers, we inserted a paraffin layer in front of the ionization chamber and were able to observe the recoil protons from neutrons. When deuterons became available for bombardment, we observed neutrons in large intensities from essentially every target used. This first observation of neutrons [9] was in September 1933.

We had frustrations—repairing vacuum leaks in the wax seals of the chamber was a continuing problem. As a leak detector we sprayed ether on the seals and observed a sharp increase in current in the ionization gauge; the laboratory often smelled like a hospital or chemical lab. The ion-source cathode was another weak point and required frequent replacement. And sometimes Lawrence could be very enthusiastic! I recall working till midnight one night to replace an ion-source filament and reseal the chamber. The next morning I cautiously warmed up and tuned the cyclotron to a new record beam intensity. Lawrence was so pleased and excited when he came into the laboratory that morning that he jubilantly ran the filament current higher and higher, exclaiming each time at the new high beam intensity, until he pushed too high and burned out the filament!

We made mistakes too, owing to inexperience in research and the general feeling of urgency in the laboratory. We had observed neutrons from every target bombarded by deuterons (originally called "deutons"), although we could not measure their energy with any accuracy. We also observed protons from every target, including one group having the same range from every target. This led to the now forgotten mistake in which the neutron mass was calculated on the assumption that the deuteron was breaking up into a proton and a neutron in the nuclear field. The neutron mass was computed from the energy of the common proton group, and came out lower than the value determined by Chadwick. Later, Tuve, Hafstad, and Dahl in Washington, D.C., using the first electrostatic generator to be completed and employed for research, showed that these protons and neutrons came from different reactions, the D(D,p) and D(D,n) processes, in which the target was deuterium-gas contamination deposited in all targets by the beam. We were chagrined and vowed to be more careful in the future.

We also had many successful and exciting moments. I recall the day early in 1934 (February 24) when Lawrence came racing into the lab waving a copy of the *Comptes Rendus* and excitedly told us of the discovery of induced radioactivity by Curie and Joliot in Paris, using natural alpha particles on boron and other light elements. They predicted that the same activities could be produced by deuterons on other targets, such as carbon. Now we had a deuteron beam in use, a carbon target on the target wheel in the chamber, and a Geiger point counter and counting circuits in service at that time. We had been making 1-min runs on another target, measuring the range of the long-range alpha particles, with the oscillator on one pole of a double-pole knife switch and the counter on the second pole. We quickly disconnected this counter switch, turned the target wheel to carbon, adjusted the point-counter circuits to count electrons, and then bombarded the target for 5 min. When the oscillator switch was opened, the counter was turned on, and click-click--click---click----click. We were observing induced radioactivity within a half hour after hearing of the Curie-Joliot results. This result was reported by Henderson, Livingston, and Lawrence [10] in March 1934.

I left the laboratory in July 1934, to go to Cornell (and later MIT) as the first missionary from the Lawrence cyclotron group. Cooksey returned to stay permanently at Berkeley, and joined in Lawrence's next stage of development, which was to expand the pole faces to 37-in. diameter and to build a larger chamber which soon reached 8-MeV deuterons; this greatly improved version was reported [11] in 1936. Many other young scientists joined the group; also the first professionally trained engineers arrived, notably W. W. Brobeck and Winfield Salisbury, and brought high-quality engineering into the laboratory. The first biological experiments were started in 1935, when Dr. John Lawrence, Ernest's brother, arrived. When a mouse was bombarded with neutrons in the re-entrant counter tube on the side of the 27½-in. machine, it came out dead. This created a strong impression in the laboratory, and was the start of the radiation-control program, though it was soon found that the mouse had really died of suffocation. Neutron shielding was installed around the cyclotron, formed of stacks of 5-gallon cans filled with water. Lawrence and his brother developed the biological program rapidly, with the strong hope that neutrons could be used for cancer therapy. Lawrence obtained support for the 60-in. Crocker cyclotron,[12] which was completed in 1939 (Fig. 16). This machine was a beautifully engineered and reliable instrument, and became the prototype of scores of cyclotrons around the world. The year 1939 was also notable as the year in which Ernest Lawrence received the Nobel Prize.

And perhaps this is the place to leave my story of Ernest Lawrence and the cyclotron. The prestige of the Nobel Prize

FIG. 16. Ernest Lawrence and his laboratory staff in 1938 within the magnet frame of the 60-in. cyclotron: (first row) J. H. Lawrence, R. Serber, F. N. D. Kurie, R. T. Birge, E. O. Lawrence, D. Cooksey, A. H. Snell, L. W. Alvarez, P. H. Abelson; (second row) J. G. Backus, Marin, P. Ebersold, E. E. McMillan, E. M. Lyman, M. D. Kamen, D. C. Kalbfell, W. W. Salisbury; (third row) A. S. Langsdorf, Jr., Simmons, J. G. Hamilton, D. H. Sloan, J. R. Oppenheimer, W. M. Brobeck, R. A. Cornog, R. R. Wilson, Viez, J. J. Livingood.

gave him the opportunity to implement his dreams of even larger cyclotrons and made him a scientific figure of international importance. The Lawrence Radiation Laboratory, with its 184-in. synchrocyclotron and the 6-GeV bevatron, is a continuing memorial to this great innovator of science.

[3]

Synchronous Accelerators and How They Grew

The principle of synchronous or phase-stable acceleration which was introduced in 1945 changed the course of accelerator history and opened a new field of scientific research, that of particle physics. The principle of phase stability was discovered independently and nearly simultaneously in two countries. In early 1945 V. Veksler in the USSR submitted a paper [1] presenting the principle. And later in 1945, before Veksler's paper reached the United States, the same principle was announced by E. M. McMillan [2] at the University of California. Both papers presented methods by which particles in resonance-type accelerators could be kept in resonance with the radiofrequency fields indefinitely, and could be accelerated to much higher energies.

The early accelerators used in nuclear research before World War II are all limited in maximum energy. Electrostatic generators and Cockcroft-Walton machines are restricted by physical limitations such as the breakdown of insulation in high electric fields. In the cyclotron the relativistic increase in mass of protons and deuterons destroys the validity of the resonance principle for energies above about 25 MeV. The betatron, although capable of reaching the 300-MeV range, is limited by the rapid increase in radiation loss by the orbiting electrons at higher energies. None of these machines can maintain acceleration indefinitely.

In resonance acceleration, particles traverse an alternating (rf) electric field applied across a series of gaps, and acquire an increment of energy on each successive traversal of this field. Synchronous accelerators use a type of phase stability possessed by certain particle orbits, so that particles having the incorrect phase or the wrong energy on arrival at the gap will experience an automatic correction toward the proper phase and energy at succeeding gaps. Particles with small phase or energy errors will continue to be accelerated, with minor oscillations in phase and energy around the correct values, and these oscillations are damped to smaller amplitude as acceleration continues. The significant feature of synchronous accelerators is that, in principle, acceleration can be continued indefinitely. At present, the upper energy limits are set only by economic considerations.

TABLE 1. Characteristics of synchronous particle accelerators.

Type	Magnetic field	Frequency of electric field	Radius of orbit
Electron synchrotron	Increasing	Constant	Constant
Synchrocyclotron	Constant	Decreasing	Increasing
Proton synchrotron	Increasing	Increasing	Constant
Linear	Zero	Constant	Infinite
Microtron	Constant	Constant	Increasing

The stability principle has had important applications in five types of accelerator: electron synchrotrons, synchrocyclotrons, proton synchrotrons, microtrons, and linear accelerators. These several types differ in the combinations and geometric arrangements of magnetic and electric fields; they also differ in the shape of particle orbits, type of magnetic field and frequency of the accelerating rf field. Their special features are summarized in Table 1.

ELECTRON SYNCHROTRON

We shall illustrate phase stability with the electron synchrotron, one of the simpler types of synchronous accelerator. In the synchrotron the electrons are retained in circular orbits of approximately constant radius in a ring-shaped transverse magnetic field. Acceleration takes place at one or more gaps around the orbit, across which a radiofrequency electric field is applied which has a frequency equal to the orbital frequency of the electrons (or an integral harmonic of it). Orbital frequency is essentially constant for electrons having a velocity approaching that of light, that is, for energies above a few million electron volts. As electron energy and momentum increase, the magnetic field must increase to maintain constant orbital radius; this is accomplished by pulsing the magnet cyclically from low to high field intensity. The electrons are accelerated in a succession of bursts at the cycling frequency.

In synchronous acceleration the electrons must acquire just the right energy per turn (on the average) to maintain constant orbital radius in the rising magnetic field; this can be called the "equilibrium" energy increase per turn. Consider an electron crossing the gap at a phase of the rf field when the potential has just the equilibrium value. There are two phases on the rf wave at which this is true, one when the voltage is rising and one when it is decreasing. The stable phase in the synchrotron is that for decreasing voltage. If the applied frequency is equal to the orbital frequency and the electron has the equilibrium energy, it will arrive at the gap in successive turns at the same phase and will stay in resonance.

Now consider a particle that arrives early (at too large a phase angle) and receives more than the equilibrium increment in energy; it will traverse a circle of larger radius than the equilibrium orbit and take a longer time than the equilibrium particle. This particle will therefore arrive at the next gap when the phase of the applied radiofrequency has shifted to a smaller value, will receive a smaller energy increment, and will traverse an orbit of smaller radius. This process continues until the energy of the particle is reduced below the equilibrium value for resonance, when the opposite response occurs. So the particle oscillates in phase of crossing the gap, its energy oscillates around the equilibrium value, and the individual orbital radii oscillate around the radius of the equilibrium orbit (Fig. 17). A similar argument shows that particles arriving

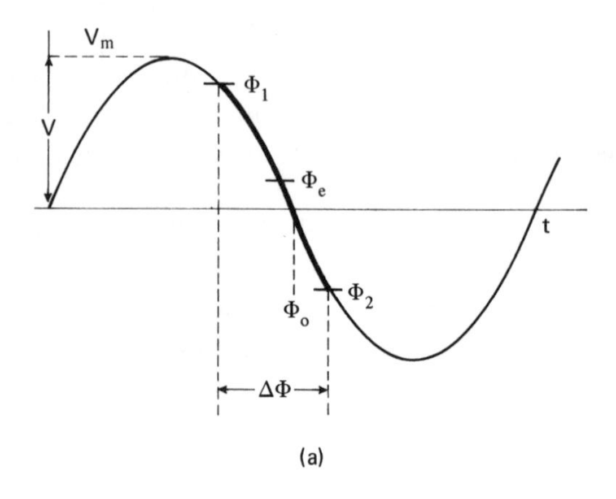

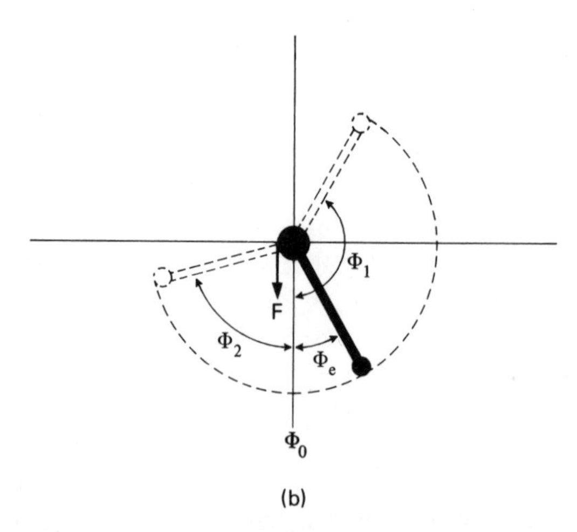

FIG. 17. (*a*) Diagram of the radiofrequency cycle of a synchrotron, in which the phase oscillates around an equilibrium phase angle Φ_e during acceleration; (*b*) analogue illustrating the motion of a displaced pendulum oscillating about its equilibrium angle of displacement Φ_e.

at the gap with the wrong energy will also oscillate in phase and in energy around the equilibrium values. Furthermore, the equations of motion show that these oscillations are damped to smaller amplitude as particle energy increases. Acceleration continues until the magnetic field reaches its maximum, which determines maximum energy. Note that the other phase of the rf, when voltage is rising, is unstable. An electron crossing the gap slightly off this phase will have its phase mismatch amplified and it will be lost from resonance. So the particles become bunched tightly around the equilibrium phase and, as the motion is damped, occupy a smaller fraction of the phase angle and have smaller azimuthal and radial spreads. This oscillation in phase and energy about the equilibrium values is analogous to the hunting in phase of the armature of a synchronous electric motor, which led McMillan to the choice of the name "synchrotron."

A virtue of the electron synchrotron is that the energy radiated by the electrons owing to their transverse acceleration in the magnetic field is automatically compensated by phase shifts to larger phase angles and higher values of volts-per-turn.[3] (The radiation loss reaches 4.5 MeV per turn in the Cambridge Electron Accelerator synchrotron, for example, with no reduction in the phase stability.)

Seldom has a new scientific principle been exploited with such promptness. The reason was that it was announced just at the end of World War II when scientists returned to their laboratories from their wartime assignments, eager to resume research activities and equipped with new skills and experience. Many new technical devices and materials were available, which had been developed during the war in fields such as microwave radar, electronics, and nuclear physics. Experience in wartime crash programs was carried over to speed up accelerator developments. But most significant was the increased national prestige of scientists, which brought prompt and generous financial support from their governments.

Unlike earlier accelerators, most of which required slow development, starting with small sizes at low energies, the synchronous machines were conceived in their full stature as high-energy devices capable of producing mesons. The first electron synchrotron to be started, by McMillan at the University of California, was designed for 320 MeV; it came into operation in January 1949. Meanwhile, in 1946, before the California machine could be completed, Goward and Barnes [4] in England demonstrated the validity of the synchronous principle with a much smaller magnet using an old betatron, in which the magnet excitation was made to over-run the betatron phase and acceleration was continued to higher energy (8 MeV), a radiofrequency circuit being used to provide the accelerating field. The next to succeed, in 1947, was a group at the General Electric Co. Research Laboratory [5] with a 70-MeV electron synchrotron, based on their previous experience with betatrons. Synchrotron radiation was first observed with this machine, as a forward-directed beam of white light.

The first wave of construction in this country was in the 300- to 350-MeV energy range. In addition to McMillan's machine at the University of California, synchrotrons were started in the years before 1949 at Massachusetts Institute of Technology (Fig. 18), the University of Michigan, Purdue University, and Cornell University; most of these were supported by the United States Office of Naval Research. At the General Electric Research Laboratory a 300-MeV synchrotron was built using a magnetic field produced by electrical coils without an iron core. Also, during this phase, a 140-MeV synchrotron was built at Oxford, England. Between 1946 and 1949 several theoretical studies of orbit stability and phase stability were published,[6] and the dynamics of particle motion were well understood. This period was climaxed by the completion in 1954 of a 350-MeV synchrotron at the University of Glasgow, which incorporated all the best features of the earlier machines.

More recently the urge for still higher energies started an-

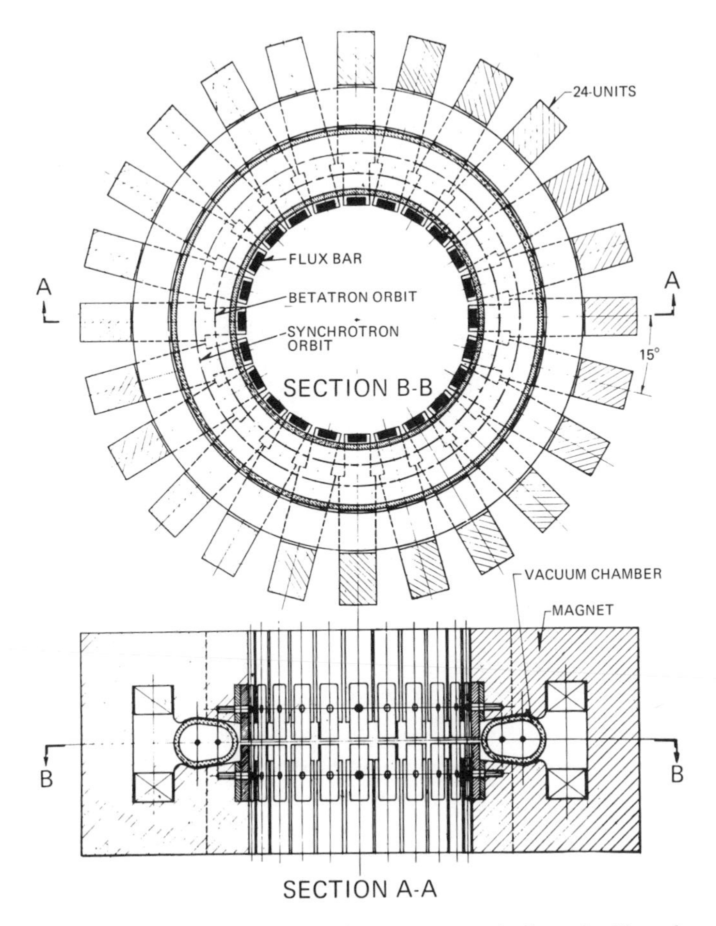

FIG. 18. Ring-shaped electron synchrotron magnet built at the Massachusetts Institute of Technology in 1948. (From *Particle Accelerators* by M. S. Livingston and J. P. Blewett. Copyright 1962 by McGraw-Hill Book Company. Used by permission of McGraw-Hill Book Company.)

other wave of construction in the 1000-MeV (1-GeV) energy range. At California Institute of Technology the magnet for the quarter-scale model of the bevatron (see below) was rebuilt by 1958 and adapted for acceleration of electrons to 1.1 GeV. At Frascati, Italy, a 1.2-GeV machine was completed in 1959.

In other laboratories the principle of alternating-gradient focusing was adopted, and machines were built at Cornell for 1.2 GeV, at Bonn University in Germany for 0.5 GeV, at the Royal Institute in Stockholm for 1.2 GeV, and at the University of Tokyo for 1.3 GeV. Still higher energies were to come, starting with the Cambridge Electron Accelerator, which more fully exploited the AG focusing principle; these will be described in the following chapter.

SYNCHROCYCLOTRON

Synchrocyclotron development started even before that of the electron synchrotron, in late 1945. In this case the problem was the relativistic limitation of fixed-frequency cyclotrons, due to the increasing mass and decreasing orbital frequency of the ions. The 184-in. magnet at the University of California had originally been conceived as a giant standard cyclotron by Professor Lawrence and his coworkers. A paper by Bethe and Rose [7] in 1938 had disclosed the relativistic limitation, but Lawrence and his group were confident that with sufficient rf voltage they could push the energy up to 100 MeV. The magnet was assembled and used for other experimental purposes during the war but was not completed as a cyclotron. At the end of the war, when McMillan proposed the use of frequency modulation, it became obvious that this method would result in higher energies, and plans were made to convert the 184-in. magnet into a synchrocyclotron.

McMillan (and Veksler) proposed that the applied frequency be modulated (reduced in frequency) cyclically, to match the decreasing orbital frequency of the ions in the uniform magnetic field as mass increased. The equilibrium phase would be shifted ahead during acceleration just enough to provide the necessary accelerating voltage to maintain an increasing equilibrium energy. In this situation a stability argument similar to that for the synchrotron applies. An ion arriving at the gap too early receives too large an energy increment, traverses a larger orbit and takes a longer time, thus arriving at the next gap with a phase closer to the equilibrium phase. Again, this leads to stable oscillations about the equilibrium phase and energy, and to radial oscillations about the radius of the equilibrium orbit. Particles are bunched about the equilibrium phase, which means in their azimuthal location in the orbit. Each sausage-shaped bunch moves radially outward in the magnetic field as the applied frequency is decreased, arriving at the periphery with maximum energy at the end of the modulation cycle. The rf modulation is cyclic with a frequency of 100 Hz or more, so high-energy particles arrive at the target in pulses at this frequency.

Another technique for maintaining resonance at constant frequency in the cyclotron would be to shape the magnet poles so as to increase the field in the outer regions to compensate for the increase in mass of the particle. However, it was known that a radially decreasing field is required for focusing in the transverse plane. If the field is azimuthally uniform and increases radially, the beam will blow up axially. In 1938 Thomas [8] proposed the use of radial sectors with alternately strong and weak magnetic field, but with the average field having the desired increase with radius. He showed that additional axial restoring forces exist at the edges of the magnet sectors which compensate for the defocusing of the average field. At this time the accelerator art was in its infancy, the theory of particle orbits was not well understood by accelerator builders, and many practical problems of engineering development took precedence. As a result, the idea lay fallow for many years. It was reactivated during the war at the University of California Radiation Laboratory, under the stimulus of potential military applications of very high beam intensity, but the work was classified secret and not published until the field was declassified in 1956. By this time it was evident that Thomas's proposal was a special case of alternating-gradient focusing as applied to constant magnetic fields, and it has had its fruition in the "sector-focused" or "isochronous" cyclotrons of the present day. This is discussed in more detail in Chapter 4.

So we return to the story of the synchronous accelerators. The first test of the stability principle at Berkeley was made on the older 37-in. cyclotron magnet in late 1945, by an ingenious method of simulating the expected relativistic mass change in the 184-in. machine with an exaggerated radial decrease in the magnetic field of the 37-in.[9] The orbital frequency decreases at high energy owing both to the relativistic mass increase and to the slight radial decrease needed for axial focusing. It can be made to decrease further in a small cyclotron if the magnetic field at large radii is reduced still further below the central field. For this test the 37-in. pole faces were tapered radially to produce a 13-percent decrease in field and so in orbital frequency; thus the techniques of frequency modulation of the radiofrequency could be studied and the principle of phase-stable synchronous acceleration could be tested. A rotating capacitor in the rf circuit was used to modulate the frequency. The results were completely successful. Very low D voltages were required as compared with conventional cyclotron operation, and the deuterons remained in resonance for many thousands of revolutions.

This success justified conversion of the 184-in. machine to utilize the principle. A crash program of development and construction resulted in initial operation in November 1946 (Fig. 19). It was successful in producing good intensities of deuterons of 190 MeV and He^{++} ions of 380 MeV. The eight authors listed in the first report [10] were leading members of the Berkeley laboratory staff; many others were also involved.

FIG. 19. The 184-in. synchrocyclotron at the University of California soon after completion in 1946 and before shielding was added. (From *Particle Accelerators* by M. S. Livingston and J. P. Blewett. Copyright 1962 by McGraw-Hill Book Company. Used by permission of McGraw-Hill Book Company.)

This was an example of the "big-team" accelerator developments that became standard in the years to follow. The impressive array of scientific and engineering talent was one reason that such a tremendous job could be completed in less than a year. But another reason was the basic soundness of the principle of phase stability and the relative simplicity of the techniques required. Equivalent success and speed in tuning up a synchrocyclotron have been achieved subsequently in other laboratories.

The immediate success at the Berkeley laboratory led others to build similar accelerators. In the United States, most of them were supported by the Office of Naval Research. Within a few years five synchrocyclotrons were completed and in operation, at Rochester, Harvard, Columbia, Chicago, and Carnegie Institute of Technology. Other installations were built abroad, at Amsterdam, Harwell, Montreal (McGill), Uppsala, Liverpool, Geneva, and in the USSR. The 184-in. machine was rebuilt in 1957, achieving an energy of 720 MeV, which is the record energy for cyclotrons; other large synchrocyclotrons are at CERN, Geneva (600 MeV), Dubna, USSR (680 MeV), and the NASA laboratory near Norfolk, Virginia (600 MeV).

PROTON SYNCHROTRON

The proton synchrotron is the culmination of phase-stable accelerators and has produced the highest energies. The synchrocyclotron requires a solid-core magnet; at relativistic energies the weight and cost of the magnet increase roughly as the square or cube of a magnet dimension such as pole-face diameter; power cost also increases about as the square of the diameter. For energies in the GeV range the weight and cost of the magnet would become exorbitant. The obvious method of reducing magnet cost is to use a ring magnet covering only a narrow annular band, and to use a pulsed magnetic field. In both McMillan's and Veksler's papers the possibility of acceleration of protons in a ring-shaped magnet was implicit, although neither paper was primarily concerned with this more complicated application of the principle.

Actually, the first proposal of a proton accelerator using a ring magnet was made in 1943 by Prof. M. L. Oliphant of the University of Birmingham, to the British Directorate of Atomic Energy. Because of wartime activities and security restrictions the proposal was not published or acted upon at that time. This antedated the discovery of phase stability by two years. But at that time there was no theoretical assurance that the proposal was sound or technically practical. In 1947, following the McMillan and Veksler papers, a theoretical analysis of orbit stability based on their work and a more detailed design study were published [11] by the Birmingham group. An accelerator following these designs was built and was in operation at 1-GeV energy from 1953 (a year later than the cosmotron) to 1967 when it was closed down. The Brimingham machine lacked some of the advantageous features developed at Brookhaven and Berkeley and operated at rather low beam intensity.

A fixed orbital radius requires synchronous acceleration. But unlike electrons, which approach the velocity of light at relatively low energies and so have essentially constant orbital frequency with increasing energy, protons do not approach this velocity until they reach energies of many giga electron volts. So the velocity and orbital frequency of protons increase during the entire acceleration interval. This requires modulation of the frequency of the applied rf field, which must synchronize with the increasing orbital frequency. This modulation must extend over a wide range of frequency, from a low value at injection of the protons to a high value at maximum energy. The rate of change of frequency must also match the rate of rise of the magnetic field, which depends on the characteristics of the power supply and the properties of the magnet iron. This introduces new and complicated technical problems in the design of the rf accelerating units and of the high-frequency oscillator, which are unique to the proton synchrotron.

This extension of the principle of synchronous acceleration was obvious to many designers, once the principle had been announced and success had been achieved with the synchrotron and the synchrocyclotron. Design studies were started early in 1947 in two laboratories supported by the United

FIG. 20. The 3-GeV "cosmotron" at Brookhaven National Laboratory, before shielding was added.

States Atomic Energy Commission, at the University of California and at Brookhaven. At Berkeley, W. M. Brobeck [12] made a preliminary study of a possible design for 10-GeV protons, which was primarily a comparison of alternative types of power supply for such a large ring magnet. Designs for a similar accelerator were begun at Brookhaven [13] under my direction, stimulated by Professor I. I. Rabi of Columbia University; the staff members involved here J. P. Blewett, G. K. Green, L. J. Haworth, and myself (on leave from MIT).

Both of these design studies were supported and encouraged by the Research Division of the AEC. When preliminary designs and cost estimates became available in 1948, a decision was made by representatives of the AEC and of the two laboratories for the construction of two machines, a 2.5- to 3.0-GeV "cosmotron" at Brookhaven and a 5- to 6-GeV "bevatron" at Berkeley. In both laboratories teams of scientists and engineers were rapidly assembled to complete designs and proceed with construction. The Brookhaven cosmotron [14] was the first to be completed, in 1952, at 2.3-GeV energy; it was brought to its maximum design energy of 3.0 GeV in 1954 (Fig. 20). The Berkeley designers chose to build first a quarter-scale model,[15] and then continued with the full-scale machine, which was brought into operation at 5 GeV in 1954 and reached 6.2 GeV in 1955. These two machines have contributed greatly to the field of particle physics.

Three other proton synchrotrons have been based on cosmotron design and experience: the 3-GeV "Saturne" at the Saclay laboratory of the French C.E.A., a rapid- cycling "PPA" machine for 3 GeV at Princeton, and a 7-GeV "Nimrod" machine at the Rutherford Laboratory in England. At Dubna, USSR, a 10-GeV "synchrophasotron" based on the bevatron design and experience was completed in 1957. And at the Argonne National Laboratory a 12.5-GeV machine called the zero-gradient synchrotron (ZGS) was finished in 1962. In the ZGS the bending magnets have uniform fields, and axial focusing is obtained by shaping the end faces of the eight magnetic sectors.

LINEAR ACCELERATORS

The earliest proposal for an electron linear accelerator was made by G. Ising [16] in Sweden in 1925. He suggested resonance acceleration of electrons down a linear array of tubular electrodes of increasing length, using a spark-gap oscillator and transmission lines to supply the radiofrequency fields of the proper phase to the electrodes. But Ising did not build a working model. This speculation prompted Wideröe to perform the first experiment in which radiofrequency resonance was observed, in an elementary linear accelerator, as described in previous chapters. And this idea of Wideröe's, in turn, led Ernest Lawrence to develop his concept of the magnetic resonance accelerator, or cyclotron.

During the early 1930's several resonance linear accelerators were built and tested. J. W. Beams [17] at the University of Virginia, in 1934, experimented with a traveling-wave accelerator

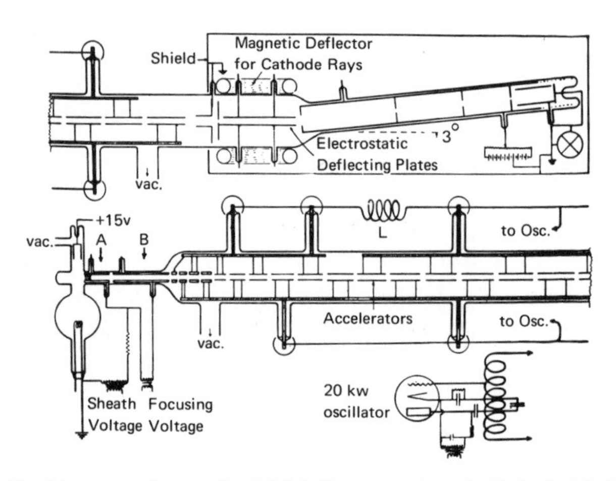

FIG. 21. Linear accelerator for 1.2 MeV mercury ions built by D. H. Sloan and E. O. Lawrence at the University of California in 1931.

for electrons using transmission lines of different lengths attached to a linear array of tubular electrodes and fed with potential surges generated by a capacitor–spark-gap circuit, similar to the system proposed by Ising. Bursts of electrons were occasionally accelerated to 1.3 MeV, but the device was not sufficiently steady or reliable for nuclear experiments.

While I was working under Lawrence at the University of California developing the first cyclotron, another graduate student, David Sloan, was building a linear accelerator (linac) with ten or more tubular electrodes in line connected alternately to a source of radiofrequency potential (Fig. 21). Adequate radio power tubes for high frequencies were not available at that time, so rather low frequencies were used (30 MHz) for which resonance could be obtained only with heavy ions.

By 1931 Sloan and Lawrence [18] had accelerated mercury ions to 1.25 MeV. Working with Sloan, W. M. Coates [19] built a linac for Hg^+ ions of about 2.8-MeV energy and B. B. Kinsey produced 1.0-MeV Li^+ ions. These heavy-ion linacs were not capable of producing significant nuclear disintegrations, and the effort was not continued.

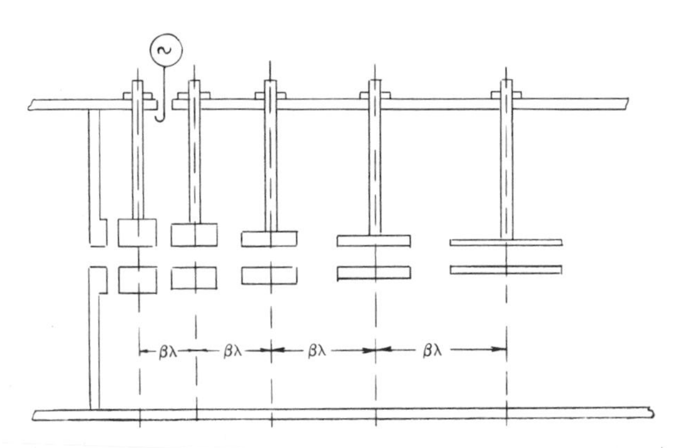

FIG. 22. Basic structure of positive-ion linear accelerator. As the drift-tube length increases, the outer diameter decreases to maintain resonance. (From *Particle Accelerators* by M. S. Livingston and J. P. Blewett. Copyright 1962 by McGraw-Hill Book Company. Used by permission of McGraw-Hill Book Company.)

Proton Linacs. The development of the proton linac had to wait for 12 years, until the improved technology in radio and radar developed during World War II made it possible. It was initiated by Luis W. Alvarez at Berkeley in 1946. While Lawrence and his associates were building the 184-in. synchrocyclotron and McMillan was engaged on his electron synchrotron, Alvarez, with the able assistance of W. K. H. Panofsky, built the first proton linac to attain an energy sufficient for nuclear experiments (32 MeV). The availability of surplus radar equipment, including rf power tubes for 200 MHz, speeded the Berkeley development. The basic structure is a long, cylindrical cavity, with drift tubes of increasing length mounted along its axis (Fig. 22). It is supplied by protons of 4-MeV energy from a Van de Graaff generator. The cavity operates in a standing-wave mode in which axial rf fields between the drift-tube electrodes provides acceleration. In this structure an ion with the correct phase sees only accelerating fields when it crosses the gaps; while the fields are in the reverse direction, the ions are shielded by the tubes. The drift tubes are basically cylinders, with rounded ends, increasing in length and decreasing in outer diameter, supported on slender stems along the axis of the cavity. The precise shapes and spacings are determined by calculations and model studies so that each "unit cell" has the same frequency. Power is fed into the cavity at several points from power tubes which are excited in parallel.

The Alvarez design has been the basis for essentially all subsequent proton linacs, and the same rf frequency of about 200 MHz has been retained. Linacs of this type are used as injectors for higher-energy machines; 50-MeV linacs provide protons for the ZGS at Argonne (Fig. 23), the AGS at Brookhaven, and the CPS at CERN. The highest-energy machine at present is the 100-MeV linac used as an injector for the proton synchrotron at Serpukhov, USSR. Construction is now in process in several laboratories for linacs to produce 200-MeV protons, and at Los Alamos construction has started on a "meson-factory" linac for 800 MeV.

These proton linacs utilize a simple type of phase stability. No magnetic fields are used and particles are accelerated in a linear path by an rf electric field of constant frequency. The separation between accelerating gaps increases along the path in such a way that the time required for a particle to traverse each space is the same, usually a full period of the rf field. The equilibrium phase is on the rising side of the voltage wave applied across the gaps. If a particle arrives too soon it will receive less than the equilibrium increment in energy, will have too small a velocity, and will arrive at the next gap at a later phase which is closer to the equilibrium phase. The reverse is also true. So the phase of crossing the gaps oscillates around

FIG. 23. The 50-MeV proton linac used as an injector for the ZGS machine at the Argonne National Laboratory.

the equilibrium phase, and particle energy oscillates around the steadily rising equilibrium energy.

Electron Linacs. Modern linear electron accelerators are based on the work of W. W. Hansen and his associates at Stanford, starting in the 1930's. One of the early concepts was a high-Q cavity resonator for single-stage acceleration of electrons. To excite this cavity, rf sources of unusually high power and high frequency were needed. This requirement led to the invention and initial development of the klystron by Hansen (in collaboration with the Varian brothers), which has since played an important role in the rf power field. At the end of the war, Hansen, E. L. Ginzton, and others started a double-headed development program, to develop waveguides for acceleration of electrons at microwave frequencies, and to provide klystrons of adequate power at these frequencies.[20] This program, ably continued after Hansen's death in 1949 by Ginzton and later by Panofsky, has resulted in a sequence of electron linear accelerators of increasing length, power, and output energy.

The accelerating structures for electron linacs are quite different from those for protons. On account of the higher electron velocities, beam apertures can be smaller and resonant frequencies higher than for protons. Largely owing to the availability of radar components in the 10-cm range, the frequency of 2855 MHz has been chosen for most electron linacs. The structure used is an iris-loaded waveguide, of about 4-in. diameter, in which a traveling wave of the TM_{01} mode provides an axial electric field with a wave velocity equal to that of light. The iris diaphragms have 2-cm apertures for the beam and are spaced along the cylindrical waveguide at intervals of about 3 cm to provide the correct wave velocity.

Electrons approach the velocity of light at an energy of a few million electron volts, so the mechanism of phase focusing is applicable only during the early part of acceleration. However, although the phase is no longer oscillating around the equilibrium value at higher energies, it now approaches it asymptotically; the electron continues to be "locked" to the accelerating wave. The electrons can be considered to ride on the advancing electric wavefront much as a surfboard rides on the front of a water wave.

A large fraction of the development of the electron linac has taken place at Stanford, and most of it has been reported in the literature, in particular the "Linear Accelerator Issue"[21] of the *Review of Scientific Instruments* (February 1955). This notable development at Stanford has culminated in the present "SLAC" 2-mile linac, which was completed in 1966 and operates at 20 GeV.

Microtron. The principle of the microtron is similar to that of the cyclotron, but it is designed to accelerate electrons rather than positive ions. It was proposed originally by Veksler.[22] The accelerating cavity is small and is near the outer boundary of the magnetic field. The successive circular orbits of the electrons are all tangent to one another within the cavity. Because the electrons very quickly attain nearly the speed of light, they gain energy and momentum but practically no speed in passing through the accelerating cavity. Each successive orbit is longer than the preceding one by one wavelength of the accelerating electric field, and the sequence of orbits forms a series of tangent circles of uniformly increasing diameter. An advantage of the microtron is the ease with which the beam can be ejected from the largest, final orbit.

A type of phase stability exists in the microtron, based on stable oscillations in the phase of crossing the accelerating cavity. A particle which arrives behind the equilibrium phase gains less than the synchronous energy, traverses a smaller circle and takes less time, and thus arrives at the next gap closer to the equilibrium phase. Although the number of accelerations is small (typically 10 to 20), some phase focusing results.

[4]

The Story of Alternating-Gradient Focusing

Progress in nuclear physics and particle physics has come in large measure from the use of a sequence of particle accelerators of higher and higher energy. The beams of charged particles which they produce, and their secondary radiations, can disintegrate nuclei and create many new types of particles. The energies achieved with accelerators have increased at an almost exponential rate during the past 35 years, averaging a tenfold increase every 6 years (see Fig. 34). The energy record has been held in turn by voltage multipliers, cyclotrons, betatrons, synchrotrons, synchrocyclotrons, proton synchrotrons and—most recently—alternating-gradient synchrotrons.

The latest in the sequence of improved concepts in the design of accelerators has been the use of alternating gradients in the guide-field magnets of synchrotrons, to provide focusing for the accelerated particles. This development has led to several new types of synchronous and resonance accelerators, capable of either much higher energies or much higher intensities than earlier machines. This principle of "strong" focusing was first publicly announced at the Brookhaven National Laboratory in the summer of 1952.

A "gradient" magnetic field is stronger on one side than on the other; it can be formed by tilting the magnet pole faces. A uniform gradient is one in which the rate of increase (or decrease) of the field is constant across the magnetic aperture; this can be achieved if the pole cross section has the shape of a rectangular hyperbola. In such a field particles entering on parallel paths will either converge or diverge; they will be either focused or defocused, depending on the sense of the gradient. Alternating-gradient (AG) focusing utilizes a sequence of focusing (F) and defocusing (D) sectors, spaced by field-free (O) regions.

A basic property of such gradient fields is that a region which is convergent (F) in one transverse plane (say the horizontal) is divergent (D) in the other transverse plane (the vertical), and vice versa. So a repetitive sequence of F, O, and D sectors will result in net focusing in both transverse planes, over a rather wide range of gradients, sector lengths, and field strengths. If the magnet sectors form a closed circular orbit, the particles will oscillate about a central orbit and be retained within the aperture of the uniform gradient fields. The frequency of these "betatron" oscillations is higher and the wavelength shorter than in the more weakly focusing magnets of earlier cyclotrons or betatrons. Amplitudes of particle oscillations about the central orbit are smaller. The magnets and the enclosing vacuum chambers of a synchrotron can be made physically smaller.

At the time when the strong-focusing principle was conceived, plans were being made for proton synchrotrons in the energy range up to 10 GeV. The reduction in magnet cross section offered by the AG principle would reduce costs for the magnets and their power supplies by a significant factor. It became economically practical to design accelerators for much larger orbits and so for higher energies.

BROOKHAVEN, 1952

I was fortunate to have had a direct association with the conception and development of the alternating-gradient principle at Brookhaven, starting in 1952. It now appears useful to record my memory of the history of this development, and to identify the sequence of events and the major contributors.

As general background, I should note that I served as chairman of the Brookhaven Accelerator Department, on leave from Massachusetts Institute of Technology, from the date Brookhaven was organized in 1946 through the period when the design of the cosmotron was essentially complete in 1948, when I returned to MIT. But I retained a strong professional interest in the successful completion of the cosmotron. Early in 1952 the cosmotron was brought into operation for the first time. I made plans to return to Brookhaven for the summer, bringing a graduate student and some meson-detection instruments, to participate in the initial research program at the cosmotron. However, the new machine could not be completed for research operations during the summer, but required engineering consolidation of the control system and modifications of some of the components.

Before I arrived at Brookhaven I had learned of the impending visit of a delegation of scientists representing the newly organized European laboratory at Geneva now known as CERN, to assess the 3-GeV cosmotron as a model for a 10-GeV accelerator. They planned also to visit the University of California to study the bevatron, which was under construction.

In anticipation of this visit by the CERN delegation, I felt it would be useful to review the design features of the cosmotron, to see if it could be extended to 10 GeV. As a start, I set myself the task of improving the efficiency of the magnet. I had been largely responsible for choosing the C-shaped cross section for the yoke of the cosmotron magnet, and was sensitive to criticisms based on its limitations. It was known that magnetic saturation effects reduced the useful radial aperture seriously at high fields, to about one-third of the width at injection fields.

In the standard synchrotron, orbit stability is achieved by use of a magnetic field with a small radial decrease, commonly expressed in terms of the "n-value," where n is the exponent of the variation in the radially decreasing field: $B_z = B_0(r_0/r)^n$. For stability of the orbit in both radial and vertical transverse coordinates in a synchrotron, the n-values must lie between zero and unity. The value chosen for the cosmotron was $n = 0.6$; this required a field that decreased by about 6 percent across the 30-in. width of the radial aperture. The poles were flat and nearly parallel.

It seemed to me that the asymmetric saturation of the C-shaped cores could be compensated, and yet the advantages of the C-shape retained, by alternating the locations of the back yokes of the magnets from inside to outside the orbit circle. However, I found that, if the pole faces were shaped to give $n=0.6$ for both types at injection fields, the asymmetric saturation at high fields would result in positive gradients in much of the gap between pole faces for one yoke location and negative gradients for the other. My first concern was whether this alternation in gradients would destroy orbit stability.

I discussed this question of orbit stability with my theoretical colleague Ernest Courant, and he took the problem home with him that evening. The next morning he reported, with some surprise, that preliminary calculations showed the orbits to be stable and to have even smaller transverse amplitudes than in the weak gradient of a standard synchrotron. As I recall, the set of gradients used by Courant in this first calculation was for $n_1=+1.0$ and $n_2=-0.2$, which was assumed to give an average value of $\bar{n}=+0.6$ as used in the cosmotron. The significance of this result was discussed with others in the laboratory, notably Hartland Snyder, and no fault could be found with Courant's analysis.

This was the start of an exciting period at Brookhaven. It seemed that if a little alternating gradient was good, more should be better. Courant's next calculations were for n-values of about ± 10, which showed even stronger focusing and smaller amplitudes of oscillation. It became clear that the average value of $\bar{n}=+0.6$ was unimportant, but that a new type of stability was associated with the alternation in gradients.

Larger and larger gradients were assumed in further stability calculations, with n values of 10, 100, 1000, and even more.

As they increased, I developed sketch designs of magnetic circuits to provide the high-gradient fields (Fig. 24). The magnet poles became more sharply tilted and narrower. We realized that the pole faces should be shaped to a rectangular hyperbola to provide a uniform gradient across the aperture. As the

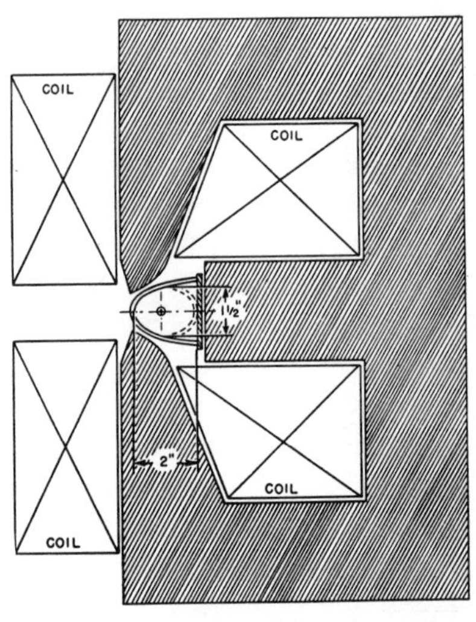

FIG. 24. Early sketch of a synchrotron magnet with very high gradients. See reference 4.2. The back-legs of the gradient magnets would be located alternately on the inside and outside of the orbit.

experimentalist on the team, I kept busy designing the strangely shaped magnetic circuits with hyperbolic pole faces and small cross sections. Some of these early speculations led to such large gradients and small apertures that construction was obviously impractical. I recall the day when the speculations reached *n*-values of 10,000, resulting in pole faces so narrow and steep and a magnetic aperture so small that the largest vacuum chamber which could be installed was less than 1 in. in diameter; at this point I objected that we had passed the bounds of practicability.

Courant also studied the synchronous oscillations in gradient fields and found them to be stable as in the normal synchrotron. Furthermore, the orbits of particles having a considerable spread in momentum were found to be compacted into a narrow radial band whose width varied inversely as the n-value. So as n-values increased and pole faces became narrower, the acceptable spread in particle momentum remained large. Stability limits were identified, and illustrated with the "necktie" diagram which became a familiar symbol of AG focusing. Suitable configurations of F and D magnets and O straight sections were devised for the arrangement of magnet units in circular orbits, such as FODO, FOFDOD, and FOFODODO.

Snyder recognized and developed the generality of the stability principle. He noted that the alternating magnetic forces on charged particles resulted in a type of dynamic stability that has many analogues in mechanical, optical, and electrical systems. For example, an inverted pendulum is unstable under static forces, and will fall to one side with any small displacement from the vertical. However, if the base is oscillated rapidly up and down through a short stroke, the pendulum is stable in the inverted position over a wide range of oscillation frequencies.

The use of gradient fields as lenses for charged particles in linear beams was studied. The magnet proposed for such applications had four poles of alternating polarity, so field intensity was zero on the axis (Fig. 25). A doublet formed of two such "quadrupoles," in which the gradients in the second unit are oriented at 90° to the first, forms a lens that focuses divergent charged particles in both transverse directions (Fig. 26). This was the first development of the quadrupole lens systems which are now commonly used in accelerator laboratories for control of linear particle beams. The strength of the focusing possible with such a quadrupole lens greatly exceeds that of a solenoidal magnetic field, and the power requirements for focusing high-energy beams are much less. We noted that similar

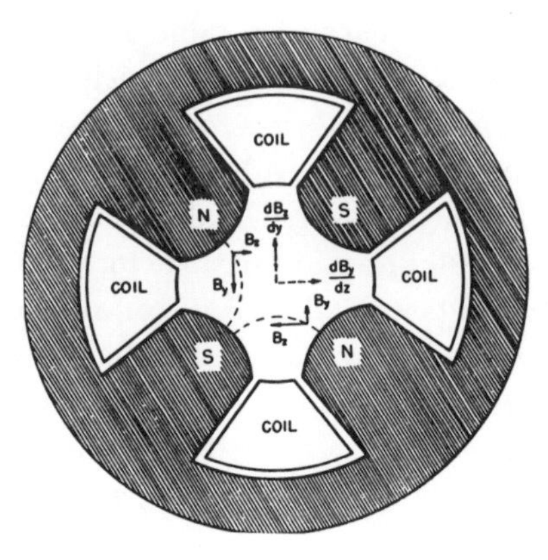

FIG. 25. Cross section of a quadrupole magnet for the focusing of linear particle beams.

focusing could be achieved with quadrupoles formed of permanent magnets. In a frivolous moment we speculated on the use of small-diameter evacuated beam pipes surrounded by permanent-magnet quadrupoles in cables many miles long, through which high-energy particle beams from the Brookhaven accelerators could be piped to the several Associated Universities that operate the Brookhaven National Laboratory.

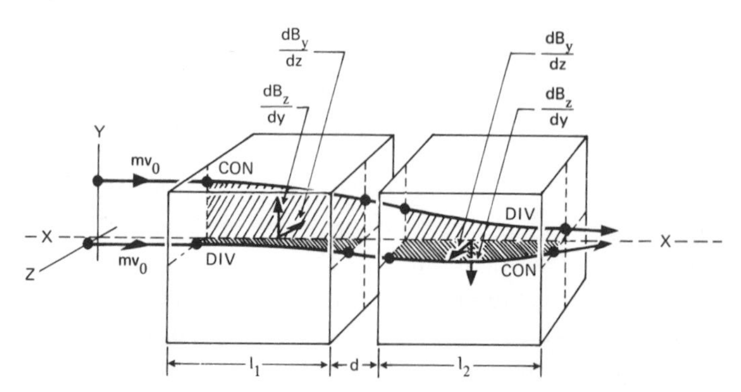

FIG. 26. Schematic diagram of the focusing of particle beams traversing two quadrupole magnets in which the directions of gradients alternate.

Meanwhile John Blewett showed that alternating electric-field gradients had the same focusing properties as magnetic-field gradients. Transverse electric-field gradients, alternating in sign, between sets of electrodes of hyperbolic cross section in a quadrupole array, will also focus particle beams along the axis. Strong-focusing electric lenses of this type appeared to be practicable for other types of accelerators. This feature was first utilized in the "electron analogue" of an AG accelerator built at Brookhaven to test the AG principle.

EARLY PLANNING FOR AG ACCELERATORS

When the CERN delegation, consisting of Odd Dahl, Frank Goward and Rolph Wideröe, arrived in Brookhaven, the AG concept had been developed sufficiently to be presented to them as a significant improvement over the cosmotron design. They were impressed with the potentialities of alternating gradients, and on their return they stimulated studies of AG orbit stability in British and European laboratories. By the time they were ready to proceed to the University of California, we realized that the development at Brookhaven had occurred so rapidly that other United States laboratories had not yet been informed. It seemed particularly important that the Berkeley group should not have to learn of this basic development through their CERN visitors. So Leland Haworth made a long-distance telephone call to Berkeley to inform them of the AG concept before the CERN delegation arrived.

Later we learned that the Berkeley staff were in the embarrassing position of being unable, owing to security classification, to describe their own developments of focusing in the cyclotron by azimuthally varying fields, as originally proposed by L. H. Thomas [1] in 1938. The use of such sector focusing in cyclotrons has since led to the development of a category of high-intensity "isochronous" cyclotrons in the energy range from 50 to 100 MeV. By the time these two lines of development merged a few years later, it became evident that Thomas's proposal of sector focusing was a special case of the general theory of AG focusing as applied to constant magnetic fields, and that the Berkeley group were also working on a type of alternating-gradient focusing at that time.

Most of these developments at Brookhaven occurred within a few weeks' time and involved primarily four staff members. The first report was sent to the *Physical Review* on August 21 and was published [2] in the December 1, 1952 issue. The report discussed the principle of alternating-gradient focusing with magnetic fields, and described design concepts for a 30-GeV accelerator. This design was a radical departure from past magnetic accelerators, using an extremely high gradient ($n = 3600$), a vacuum chamber with an aperture of only 1×2 in., and

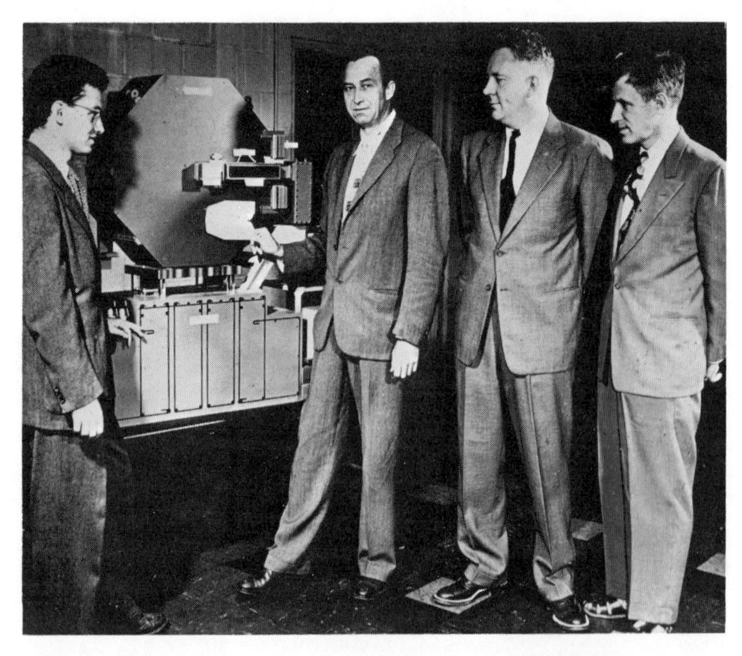

FIG. 27. E. D. Courant, M. S. Livingston, H. S. Snyder, and J. P. Blewett demonstrating the relative cross sections of the cosmotron magnet and a speculative alternating-gradient magnet of very large gradient.

an orbit circle of 300-ft radius (Fig. 27). The report also described a quadrupole magnetic lens system for focusing high-energy beams in linear systems. A companion paper by Blewett[3] presented the parallel case of focusing by electric-field gradients, and described its application to linear accelerators.

By this time it had become clear that the AG principle had important potentialities and deserved thorough study. A major step was reported by Adams, Hine, and Lawson[4] of the Harwell Atomic Energy Establishment early the next year. They identified and studied the problem of orbital resonances which might threaten orbit stability. If the frequency of the transverse "betatron" oscillations were an integral multiple of the orbital frequency, the effect of even a very small orbit perturbation would be cumulative and could build up disastrous amplitudes in a few revolutions, resulting in loss of the beam. To a lesser extent the same is true of half-integral and other subintegral resonances. For a time this seemed to set a serious limitation to the use of alternating gradients. But further work showed that objectionable resonances could be avoided by care in design and by use of suitable control systems to maintain constant nonintegral frequencies during the acceleration cycle, even with n-values as large as 100.

As so frequently happens in a technical field, this concept was developed independently elsewhere. N. C. Christofilos, an electrical engineer of American birth, educated and working in Athens, had been studying accelerators as a hobby for some years. He suggested several new and unusual ideas on accelerator design, in the form of private reports and patent applications. An unpublished report[5] dated 1950 presented the concept of AG focusing and the conceptual design of an accelerator using this principle. He also applied for United States and European patents. A copy of his report was privately transmitted to the University of California Radiation Laboratory at that time, but was not given serious consideration.

After the Brookhaven publication in 1952, Christofilos came to the United States and demonstrated his priority. This was recognized in a brief note published by Courant, Livingston, Snyder, and Blewett[6] in 1953. Christofilos joined the staff of the Brookhaven Laboratory for a time, where he continued his speculative designing of accelerators and other devices and also contributed to the laboratory program leading toward a large AG accelerator and to its linear-accelerator injector.

DESIGN STUDIES

In the United States, I started a design study late in 1952 at Massachusetts Institute of Technology, with the assistance of members of the MIT and Harvard Physics Department staffs. It resulted in a laboratory report[7] dated June 1953. This was the earliest relatively complete design study for an AG accelerator, but it was dropped when financial support for the AGS at Brookhaven was authorized by the U. S. Atomic Energy Commission in 1954. However, the Cambridge group promptly transferred their interest to electrons and designed the first multi-GeV electron synchrotron using AG magnetic focusing, which was later constructed and is now in operation as the 6-GeV Cambridge Electron Accelerator at Harvard.

Both the Brookhaven and the CERN groups initiated design studies in 1953 for AG accelerators in the energy range of 25 to 30 GeV. This energy was ten times that of the cosmotron, the highest-energy accelerator then in operation, and five times that of the bevatron, which was under construction. The collaboration between the two groups, which started in 1952, continued with exchange of staff and design information. As a result the two machines have striking similarities. Theoretical groups at both laboratories were concerned with the effects of magnet misalignments and with beam behavior at phase transition, and both groups wished to check the theory with a working model before committing the final design. A compromise acceptable to both was for Brookhaven to build an elec-

FIG. 28. The 28-GeV alternating-gradient proton synchrotron (CPS) at the CERN Laboratory in Geneva.

tron model using electrostatic-gradient fields, while CERN proceeded with detailed design. This "electron analogue" was first operated in 1955 at an energy of 10 MeV, and supported the most optimistic theoretical predictions. Construction could proceed with confidence.

The CERN group was initially under the direction of O. Dahl of Norway and F. Goward of England. Later, Dahl returned to Norway and, following Goward's untimely death, direction of the design study passed to J. B. Adams of England and C. Schmelzer of Germany. The CERN CPS was brought into operation at 26 GeV in 1959 (Fig. 28). The Brookhaven

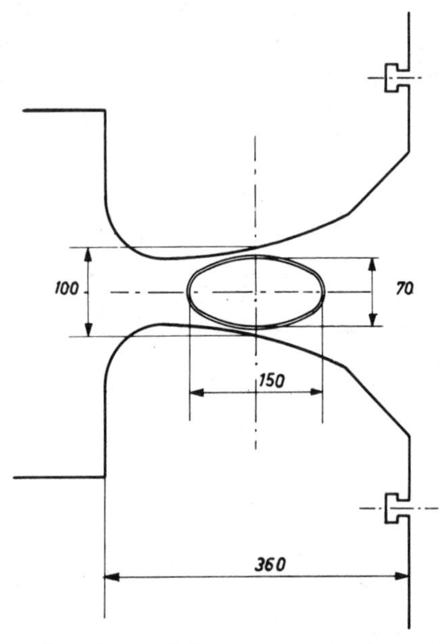

FIG. 29. Cross-section sketch of the magnet pole tips and vacuum chamber for the CERN 28-GeV CPS (dimensions in millimeters).

design group was headed by L. J. Haworth, G. K. Green, and J. P. Blewett. The AGS was completed in July 1960 and soon exceeded its design energy to operate at 33 GeV (Fig. 29). Both machines have been highly successful and each has supported an impressive program of research experiments. Beam intensities have far surpassed the designer's estimates, and the use of multiple targets and emergent beams has broadened their capabilities and effectiveness for research support.

The possibility of extension of the AG principle to fixed-field magnets (FFAG) was soon recognized, by scientists in many places, including Brookhaven, Japan, and the USSR. Another United States group formed a design laboratory called Midwestern Universities Research Association (MURA), whose primary function was to design a high-energy accelerator for the region; they emphasized the FFAG principle in their design studies. This is not the place to go into the history of MURA, but the group has made major contributions to accelerator theory and technique. The most important application of fixed-field alternating-gradient focusing up to the present is that used in "sector-focused" or "isochronous" cyclotrons, mentioned earlier as based on the 1938 proposal by Thomas and first developed at the University of California.

The first working accelerator to use the AG focusing principle was a 1-GeV electron synchrotron being constructed at Cornell University at that time by R. R. Wilson and his associates. The magnet had replaceable pole-face units intended originally to be of the uniform-gradient type. However, when the Cornell group heard of the Brookhaven developments, they rapidly designed and ordered pole-face units with alternating gradients. By late 1953 the Cornell machine was operating at 1-GeV energy. Although the AG pole tips were not essential to the operation of this synchrotron, they provided the equivalent of a larger aperture and somewhat higher intensities. The Cornell group has recently (1967) completed a 10-GeV electron accelerator using alternating-gradient focusing.

The most important application of AG focusing has been to proton synchrotrons. It has led to the CERN CPS which operates at 28 GeV, and the Brookhaven AGS which for many years held the world's record with an energy of 33 GeV. A larger machine has recently been completed in Serpukhov in the USSR, operating at 70 to 80 GeV. The AG principle has completely changed the basic designs for the magnets used for synchrotrons. Particle orbits can be retained between much smaller magnet poles and within smaller vacuum chambers. The basic principle is still that of synchronous acceleration, but magnet apertures and cross sections do not need to increase as orbit radius is made larger. It has become economically practical to design synchrotrons for much higher energies.

PLANS FOR THE FUTURE

Research scientists in the field of particle physics are clamoring for accelerators of still higher energies, and designs are in process in several countries. In all of these design studies the decision has been simply to enlarge the orbital dimension of the AG proton synchrotron. In general, the basic design concepts and technical features of the present 30-GeV machines are believed to be capable of extension to energies of several hundred GeV. The transverse dimensions of magnets and vacuum chambers do not increase with increasing energy. All the scientific and engineering problems of machines of 200- to 300-GeV energy appear to be solvable. The major problem is that of cost, which is nearly proportional to energy and is estimated at present to be about $1,000,000 per GeV for the machine and basic laboratory facilities for research utilization.

Plans are under way in the United States for an accelerator of 200-GeV energy or higher. A design study at the Lawrence Radiation Laboratory of the University of California at Berkeley for a 200-GeV machine provided the first basis for plans in the United States. The National Accelerator Laboratory at Oak Brook, Illinois, under the direction of R. R. Wilson, has prepared a revised design study and a proposal for construction for the machine, which has been submitted to the Atomic Energy Commission and the U. S. Congress. A site has been chosen at Weston, Illinois, and the staff of this National Accelerator Laboratory has been assembled. In Europe a group at CERN has submitted a preliminary design for a 300-GeV machine to their governments, and approval is anticipated. More speculative designs exist for energies up to 1000 GeV. The only known limit is cost. Scientists in the field are convinced that these new machines are necessary to further progress in particle physics and the study of nuclear forces.

It is clear that the principle of alternating-gradient focusing has played an essential part in the development of multi-GeV accelerators and of the research field of particle physics. The reduction in unit costs that it has made possible has allowed the energy range of accelerators to be extended into a region of extraordinary scientific interest. The even higher-energy machines of the future will involve costs that will become significant items in national budgets. Their most significant product will be new knowledge about the fundamental particles of nature and the properties of nuclear forces. Justification for support of this research field must be found in the value of this new knowledge to our society. Many scientists believe that we are at the threshold of a significant upward step in human understanding. Ultimately, the impact of this new knowledge will be felt on a very broad intellectual and cultural level. High-energy physicists and accelerator enthusiasts are hopeful that the giant accelerators now being planned will be built and will contribute significantly to this growing understanding of nature.

[5]

Origins of the Cambridge Electron Accelerator

The Cambridge Electron Accelerator can be described in vector notation as the sum of three forces: the Harvard Physics Department's desire to have an accelerator capable of producing mesons, the continuing MIT interest in high-energy electron and photon physics, and my own experience in the design of accelerators. It was indeed fortunate that these vectors pointed in the same direction in the planning years of 1952–1956, so their sum was the unique and successful 6-GeV synchrotron at the CEA. The joint administration was made possible by the growing interest at Harvard and MIT in providing closer cooperation between the two institutions. Nationally, it came at just the right time to take advantage of policy decisions within the Atomic Energy Commission to support accelerator facilities in the larger universities in addition to the National Laboratories.

The CEA was the first multi-GeV electron accelerator using alternating-gradient focusing, and held the energy record for electrons from its completion in 1962 until shortly before the SLAC 2-mile linac at Stanford came into operation in 1966. It was the first synchrotron to take full advantage of the small apertures possible with AG focusing, so its magnets were small, compact, and relatively inexpensive. The magnet cycling rate of 60 Hz was achieved with a unique resonant powering circuit of modest power level. It utilized an unusually high radiofrequency for acceleration and phase focusing, and high rf power for compensation of radiation losses. The accelerated beams were reduced by radiation damping to such small dimensions and energy spread that they could be ejected and focused onto external targets on a spot a few millimeters across, a tremendous experimental advantage.

Long before the CEA was completed, its unique design features and the growing scientific interest in electron and photon physics made it the model for developments in other laboratories. The Deutsches Elektronen Synchrotron Laboratory in Hamburg, Germany, sent a sequence of staff members to Cambridge to learn-by-doing at the CEA; this "DESY" machine was completed in 1965, and has been run at energies up to 7.0 GeV. The "NINA" machine at the Daresbury Laboratory (near Liverpool, England) was completed and dedicated in 1967; it operates at 4.5 GeV. At Cornell University a large-orbit, somewhat simplified AG synchrotron rated for 10 GeV was brought into preliminary operation in 1967. And at the Physical Institute of the Armenian Academy of Sciences in Erivan, a 6.5-GeV machine based on the CEA called "ARUS" was completed in 1967. So the CEA has had an impact far outside Cambridge, its staff members have earned the respect of accelerator scientists throughout the world, and it has "put through school" a long list of younger "sisters."

The Harvard interest which culminated in the CEA can be traced back to about 1948, even before the synchrocyclotron which is just terminating its career was completed, when the production of π-mesons by the 184-in. cyclotron at Berkeley showed that the 150-MeV energy of the Harvard machine was just below the meson threshold. The cyclotron was built by K. T. Bainbridge, R. R. Wilson, W. Davenport, and others, and a research program was initiated, but the Harvard group anticipated the need for, and kept alert to the possibilities of, a higher-energy machine. The 1952 Report of the Visiting Committee to the Physics Department (J. Robert Oppenheimer, chairman) said: "We believe it not unlikely that within the next five years Harvard will desire to build at least one new major piece of equipment to fill this gap." So in 1952 the Harvard physics group were actively searching for an opportunity to supplement the cyclotron, when the first discussions started with MIT on the feasibility of a jointly sponsored accelerator.

At MIT, the postwar choice of a high-energy accelerator was

an electron synchrotron, with an energy of 300 MeV. The principle of synchronous, or phase-stable, acceleration of electrons in a ring magnet using a pulsed magnetic field was announced in 1945, as we have seen, independently by E. M. McMillan at Berkeley and V. Veksler in the USSR. By 1946 the technique had been demonstrated experimentally and theoretical analysis of particle motion was in an advanced stage. An experimental group headed by Ivan Getting built the MIT electron synchrotron in 1946–47 and started studies on the photoproduction of mesons. This was one of the first electron machines to use a C-shaped core for the ring magnets, and the MIT group contributed several other important design innovations. By 1952 this group was deeply involved in π-meson physics, and were planning for a larger synchrotron with an energy tentatively chosen as 1.5 GeV.

Meanwhile, I returned to MIT in 1948 from my position as chairman of the Accelerator Department at Brookhaven, where I had helped initiate the accelerator program and the preliminary designs for the cosmotron. By early 1952 the cosmotron was approaching completion and I had turned my interests to the local needs in Cambridge. I joined in discussions with Jerrold Zacharias, then chairman of the MIT Nuclear Science Laboratory, and Norman Ramsey, representing the Harvard physics group, on the feasibility of an accelerator in Cambridge. We also knew that certain European scientists were planning a proton synchrotron of about 10-GeV energy, to be based on the cosmotron or the Berkeley bevatron designs. On May 14, 1952, I sent a memorandum to President Stratton of MIT proposing a summer design-study group to anticipate these needs, and approached Leland Haworth, director at Brookhaven, to release Ernest Courant, Hartland Snyder, and others for such a study.

Haworth felt that he could not release any Brookhaven staff for that critical time and made a counterproposal that I come to Brookhaven for the summer (1952), in which case their key people could devote part of their time to a design study. I accepted this offer and went to Brookhaven. One of the first results of this study, as we have seen, was the concept of alternating-gradient focusing, which was immediately recognized as of potential importance in the design of higher-energy machines. Before the end of the summer, I joined with Courant and Snyder in publishing the first report of the AG principle, including a preliminary design for a 30-GeV machine.

Meanwhile, Ramsey came to Brookhaven in August and became enthusiastic about the prospects for AG machines. We immediately started preparing a draft of a proposal to the Atomic Energy Commission for a jointly sponsored accelerator of this new type to be located in Cambridge, and discussed the possibility with Thomas Johnson, then director of the AEC Research Division, when he visited Brookhaven later that summer. On our return to Cambridge we organized a self-constituted "Joint Accelerator Committee," consisting of Ramsey and J. Curry Street from Harvard and Zacharias and myself from MIT, to carry on the planning. Although this initial committee was an ad hoc group, other department members and both university administrations were kept informed and gave us full support. On October 10, 1952, we submitted a preliminary draft, and on October 31 a more definite request to the AEC, for support of a joint design study of a 10–20-GeV AG proton synchrotron to be carried on in Cambridge under my direction. This request was approved, and arrangements were made for funds to be supplied through a subcontract from Brookhaven to MIT for the period from November 1, 1952, to June 30, 1953.

I quickly organized a design study and arranged for laboratory space at MIT. Bulletin No. 1, dated November 7, 1952, scheduled a series of six planning conferences within the next week on the systems design and the several major components. This study involved at the start at least 20 members of the MIT and Harvard departments. A small group of full-time people were enlisted, including R. Q. Twiss from England, to extend the theoretical analysis, and J. A. Hofmann, A. Vash, and J. F. Frazer, who built dc and ac models of high-gradient magnets and instruments to measure gradient fields. The result of this joint effort was a 200-page report: "Design Study for a 15-BeV Accelerator," published as *MIT Laboratory for Nuclear Science, Report No. 60,* dated June 30, 1953. This was the first relatively complete design study of a multi-GeV AG accelerator; it included theoretical studies of resonances, synchronous stability, phase transition, damping, and the like. It also included a table of parameters (for example, orbit radius 124 ft) and a cost estimate (about $8,000,000). On August 11, 1953, the Joint Accelerator Committee, acting with the approval of the university administrations, forwarded to the AEC a proposal, backed up by the design study, for the construction of this accelerator in Cambridge.

Meanwhile, the Brookhaven staff were engaged in their own design study of a larger AG accelerator, initially conceived to be for 50 GeV and later reduced to 25–30 GeV, to be located at Brookhaven. Paralleling this effort, some of the European scientists who were then organizing the CERN laboratory coordinated their design studies in several countries; this effort culminated by late 1953 in a design for an AG proton synchrotron of about 25-GeV energy.

In September 1953, T. Johnson of the AEC informed us that his office would recommend support of the Brookhaven proposal for a 25–30-GeV machine but could not support a competitive Cambridge proposal. Although we brought the heaviest guns from both university administrations to bear, this recommendation was confirmed, and we were informed in December of the Commission's decision to refuse the Cambridge proposal. This led to renewed political and technical activities in Cambridge, including discussions with the National Science Foundation as a possible alternative source of supporting funds. These discussions led the NSF to schedule a meeting of an Advisory Panel on Ultrahigh Energy Nuclear Accelerators for early spring 1954, at which Ramsey testified for the Cambridge group, and which proposed a consistent United States accelerator program.

ELECTRON SYNCHROTRON

On the technical side, the AEC decision led to further discussions between Harvard and MIT scientists, in which an electron accelerator was seriously considered as an alternate possibility. Although protons were clearly superior for studies of strong nuclear interactions, the prospects of a large proton accelerator at Brookhaven reduced the attractiveness of a lower-energy proton machine in Cambridge. Also, the MIT group working with their 300-MeV synchrotron wished to continue their electro- and photoproduction studies of mesons. For a time a hybrid machine for 6-GeV protons and 2.5-GeV electrons was considered. But soon it became clear that our best hope for a unique facility in Cambridge would be a high-energy electron machine.

I was personally convinced that AG magnetic focusing could be applied to an electron synchrotron, despite the very fast cycling rate required. The AG magnet lattice with field-free straight sections between magnet sectors was ideal for installation of an rf system having many high-Q resonant cavities to develop the large value of the volts-per-turn ratio needed to compensate for radiation losses. The fast cycling and the high source intensity possible with electrons should increase average beam current and largely compensate for the lower production cross section for nuclear processes through the electromagnetic interaction. And we hoped that theoretical interpretations might be simpler when the known electromagnetic interaction was involved. During the winter and early spring of 1954, I drew sketches and calculated parameters for a sequence of AG electron synchrotrons with increasing energies, which were discussed by the Harvard–MIT group.

As our confidence grew we realized that the choice was one of balancing the obvious advantages of very high-energy electrons against the increasing complexity and cost of such a high-energy machine. Our desire was to build a facility for university research, convenient to the scientists and students and operated as nearly as possible like a university research laboratory.

A site location at Harvard was agreed on at an early date. The filled land near MIT was clearly unsuitable for the underground-tunnel building needed for shielding. Sites outside Cambridge would be much less convenient for the university users. The Harvard administration gave tentative approval to our plans to use the area around and beyond the Cyclotron Laboratory, which was undeveloped at that time, was close to the existing Harvard physics buildings and laboratories, and was suitable for excavation. Both MIT and Harvard representatives approved of this location.

The National Science Foundation Advisory Panel on Ultrahigh Energy Nuclear Accelerators, referred to earlier, met on March 23, 1954 in Washington, and Ramsey presented arguments for accelerators in universities. The resulting Panel Report recommended federal support for some university machines in addition to those in the existing National Laboratories, and was important in establishing sound principles for support of the newly developing field of high-energy physics. It also opened the possibility that the NSF might support university accelerators if the AEC did not, which may have had some influence on AEC policy.

The group of interested Harvard and MIT scientists continued their discussion meetings during the spring of 1954, and the design of an electron accelerator took form. I summarized the status in a report, "Design Study for an AG Electron Synchrotron," April 26, 1954, which was later given the report number CAP-1 when the "Cambridge Accelerator Project" organization was established. This report gave design parameters for 6-GeV protons and 2.5- and 5.0-GeV electrons, with rough cost estimates. The Cambridge Joint Accelerator Committee used the 5-GeV electron design data from this report and recommended it to the Harvard–MIT group in an unnumbered report in May 1954. This became the clear choice of the Cambridge group. It was presented to the Harvard Nuclear Physics Committee and approved unanimously at a meeting on June 7, 1954. During the summer the design studies continued, and the energy being considered rose to 6.0 GeV. Representatives of the Joint Accelerator Committee (Livingston and Ramsey) prepared a "Proposal for a 5- to 6-BeV Electron Accelerator," dated August 17, 1954 (later given the number CAP-2), which included parameters and cost estimates, and submitted it to the AEC.

On August 18, I sent a request to Johnson at the AEC for funds to continue the design study during the interval before the proposal could be acted upon. This request was deferred from month to month, and the local design group was reduced to a few part-time people.

During this interval our understanding with the AEC Research Division was that they would recommend construction of two university accelerators, one at Cambridge and another at Princeton (later Princeton–Pennsylvania) and that we could expect Congressional action in the spring of 1955.

During the Physical Society meetings in New York in January 1955, I heard some disturbing rumors of a change in policy and planning within the AEC. When traced down they proved to be valid. Following a recommendation by the General Advisory Committee, the AEC had removed the construction items for the two accelerators from their budget, and was requesting only design funds for the coming fiscal year. Furthermore, Johnson had decided that the AEC would make a new survey of all accelerator requests throughout the country, and would support only the most promising with design funds. First priority would go to the National Laboratories, second to groups of universities, and third to individual universities. This seemed to us to be a reversal of AEC policy regarding university accelerators, and it certainly would cause a delay of a full year in our plans for Cambridge.

This delay was very disappointing to the Cambridge group. We felt it was essential to enlist an effective design staff immediately and to proceed with the detailed design. To meet the increased competition it would be necessary to submit a much more detailed proposal than our first one, but we had no funds to carry on the design study. In this emergency the Joint Accelerator Committee approached the Harvard and MIT administrations for special funds. I prepared a minimum budget of $33,000 for six months, half to be provided by each of the institutions. On May 17, E. M. Purcell and I, representing the Joint Accelerator Committee, met with administration officials to work out details, and on June 23, 1955 we received authorization for the amount needed from each institution, so we could proceed with the design study. The study was to be known as the Cambridge Accelerator Project (CAP) and would be administered by the Harvard Office for Research Contracts.

During the spring of 1955 Congress somewhat unexpectedly reinserted $10,000,000 in the AEC appropriations bill for two university accelerators. However, the AEC continued with its plans for what we called the "box-top" contest. On July 6, Thomas Johnson sent letters to the presidents of Harvard, MIT, and a large number of other universities requesting accelerator proposals, with a deadline for submission of October 1. At the request of other applicants this date was extended to January 1, 1956. M. G. White at Princeton saw the virtues of a joint proposal such as ours and revised his plans to include the University of Pennsylvania. A few other proposals were submitted.

Meanwhile, the CAP took form in Cambridge and our design activities intensified. During the summer Giovanni

Lanza, Kenneth Robinson, and Ralph Waniek were employed full time, and 20 Harvard and MIT staff members contributed part-time help. The final result of this effort was a 64-page "Proposal for a 6-BeV Electron Accelerator" (CAP-15), dated December 15, 1955, which was submitted to Johnson of the AEC by Presidents Killian and Pusey of MIT and Harvard. The CAP group continued technical studies during the following spring, producing a total of 25 reports and continuously modifying the design parameters. I note, for example, that the final (enlarged) orbit radius of 118 ft was first reported in CAP-18, January 6, 1956, and that revised parameters for the radiofrequency system were set in CAP-25, dated May 1, 1956. We found that the cost estimate in the proposal would not be sufficient to build the enlarged accelerator and also an experimental laboratory sufficient to utilize its capabilities. This resulted in a supplementary proposal for an additional $2,000,000, dated June 1, 1956.

A "letter of intent" contract between the AEC and Harvard University for funds to construct the accelerator was signed on April 2, 1956. This became the official starting date of the CEA. A letter of agreement had been exchanged earlier between the presidents of the two institutions, dated March 7, 1956, which gave it the name Cambridge Electron Accelerator, described the intent of sharing the research opportunities equally, established the administrative system of an Executive Committee with equal representation from Harvard and MIT and, since the accelerator was to be located on Harvard property, named Harvard University to be the contracting agency and to provide contractual and budgetary supervision. The total funds authorized in the original definitive contract were $6,500,000, for construction of the laboratory and the accelerator. Later proposals covered the anticipated annual cost for research operation following completion. The eventual cost of construction of the machine and the full-size laboratory was $11,600,000; this is just under $0.002/volt.

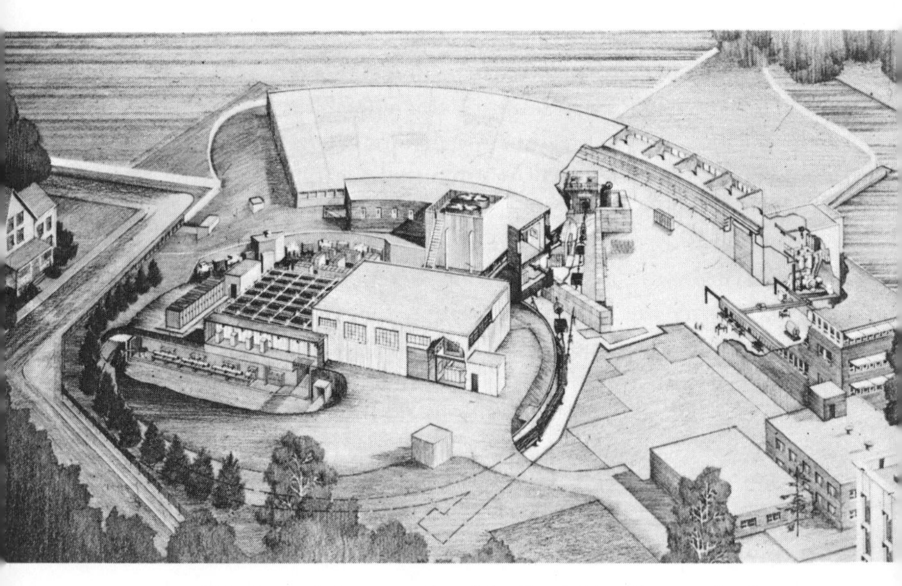

FIG. 30. Artist's conception of the Cambridge Electron Accelerator.

The first addition to the staff following the official start of the CEA was our late dearly beloved business manager, Robert E. Cummings. Lanza, Waniek, and Robinson continued from the CAP roster. Others joining us during the next year were Paul Cooper, John Rees, Lee Young, Tom Collins, Janez Dekleva, Maurice Vallauri, Manfred Wachter, and several part-time staff such as Andrew Koehler and Frank Robie. After occupying various temporary quarters, we settled into the present CEA building when it was completed in 1958 (Fig. 30).

The accelerator was completed and research activities were started in late 1962. It has performed beyond design specifications in almost all respects. One of its most valuable features is the high quality and the flexibility in control of the emergent photon and electron beams as developed by the CEA staff. The number and variety of emergent beams have provided opportunities for an unusually large number of experimental groups. The number of research scientists and students in-

FIG. 31. The magnets, radiofrequency system, and other components of the Cambridge Electron Accelerator in 1962.

volved in experiments at the CEA is over 150. The quality of the research has been consistently high and the significance of the scientific results can hardly be overestimated. It has become a successful and valuable university research laboratory (Fig. 31).

[6]

The 200-GeV Accelerator *

The 200-GeV accelerator is the highest-energy machine in the sequence described in this monograph, and is the culmination of over 30 years of scientific and technical development. It is basically an alternating-gradient proton synchrotron of very large orbital radius. The conceptual design represents the cumulative efforts of hundreds of professional accelerator scientists and engineers, including many from foreign laboratories. The origin of individual technical concepts is becoming increasingly difficult to identify, and the assignment of credit must be widely shared between many individuals. A large group of research scientists who have supported the proposal and hope to use the accelerator for experiments are broadly spread across the country. Also, much of the planning effort which has made the 200-GeV machine a reality has come from senior scientists who are not themselves experts in this field of science, and from government administrators and political leaders responsible for maintaining national scientific progress. The design staff that has been assembled has come from many accelerator laboratories and universities. The site has been chosen to make the laboratory a facility for scientists throughout the nation. It is expected to become also an important international center for research in high-energy particle physics. The project was authorized for construction by the U. S. Congress on April 19, 1968, and is scheduled for initial operation in 1972.

ORIGINS

The conceptual planning for an accelerator in the hundred-GeV energy range became possible with the discovery of the principle of alternating-gradient focusing in 1952 (Chapter 4) which led to the design and construction of the 28-GeV CPS at

CERN and the 33-GeV AGS at Brookhaven National Laboratory. Although these two accelerators represented the practical first steps toward exploiting the new principle, speculative thinking which was aimed at much higher energies started as early as 1952 in Brookhaven and at several other laboratories in the years to follow.

At the University of California Radiation Laboratory the design staff had been thinking for years about the next step beyond the 6-GeV bevatron. W. M. Brobeck made an early feasibility study of design features and methods of powering magnets for synchrotrons in the 100- to 150-GeV range in 1955. In 1956 a local Accelerator-Building Committee was formed, consisting of E. M. McMillan, E. J. Lofgren, R. L. Thornton, W. M. Brobeck, L. Smith, and D. L. Judd, which coordinated planning during the next 5 years during which nearly 50 internal notes and memorandums were prepared. This program was summarized by Judd in 1960.[1]

This early planning was brought to a focus at a summer study in 1959 of the Midwest Universities Research Association (MURA) at Madison, Wisconsin, where questions were first raised and discussed publicly concerning the scientific need and feasibility of an accelerator for energies of several hundred GeV. A memorandum [2] circulated that fall by Matthew Sands, then at California Institute of Technology, described a possible 300-GeV "cascade synchrotron" using a very large ring of alternating-gradient magnets and a smaller synchrotron as an injector. This discussion came at a time when the CERN CPS had just been brought into operation at 26 GeV and the Brookhaven AGS was approaching completion.

Even at that time it was recognized that costs would be nearly proportional to energy and that the unit cost would be similar to that at CERN and Brookhaven, about $1 million per GeV. Many realized that such a large and costly accelerator might properly be supported by an international laboratory similar to CERN. Discussions were held at the Rochester High Energy Physics Conference[3] in August 1960 with many foreign delegates attending. In September, at a meeting held at the American Institute of Physics in New York between United States and Soviet physicists, it was agreed that both countries should explore further the feasibility and desirability of accelerators for energies above 200 GeV. Plans were made to report progress at the next International Accelerator Conference, scheduled to be held at Brookhaven in September 1961.

Three United States laboratories carried on design studies during the early 1960's, at Brookhaven, California Institute of Technology, and the University of California Radiation Laboratory at Berkeley. Abroad, a design group consisting of experienced members of the CPS staff at CERN initiated studies, and later events showed that preliminary studies were also under way in the Soviet Union.

At Brookhaven, a group led by J. P. Blewett started thinking seriously about higher-energy machines in 1960. They chose energies of 400, 700, and 1000 GeV for their initial study. A preliminary design report was issued in May 1961, and conclusions were revised during an extensive study sponsored by Brookhaven during August 1961. At one time, about 25 accelerator experts were assembled from nine centers in the United States, with others from CERN and the Rutherford Laboratory in England. The Report[4] of this study group was published in late 1961; it included both a preliminary design and an analysis of experimental program requirements.

These early concepts for superenergy machines were strongly influenced by the design of the Brookhaven AGS, but even more by the belief that large gradients could be achieved, with very small magnets. The magnets designed were compact and simple. But the orbital radii were so very large that the engineering planning for other components and for powering the magnets was quite nebulous. A sequence of improved and revised tables of parameters for the three energy ranges was developed during the next few years.

The Caltech group organized by Sands made a series of studies during 1960 and 1961 and issued several internal reports.[5] They studied a variety of special problems, such as the preacceleration of the ions in a smaller "booster" synchrotron and injection into the large orbit. When it became clear that the University of California Radiation Laboratory and Brookhaven were becoming involved in similar design thinking, a meeting was held on neutral ground at UCLA in December 1961 between members of the three groups. An agreement was reached under which UCRL would carry on the West Coast design efforts in the 100- to 300-GeV range, and Brookhaven would continue to explore the higher energy range. The Caltech effort was phased out in the following year.

Design planning was followed up much more seriously at the University of California Radiation Laboratory. The status of UCRL studies was reported by Judd and Smith[6] in late 1961. In February 1962 the laboratory submitted a formal request to the United States Atomic Energy Commission for support of a design study in the 100–300-GeV range and renewed the request in December 1962. Early concepts at UCRL were similar to those at Caltech and Brookhaven in that they involved a large ring of alternating-gradient magnets for the main synchrotron. But initially they explored the use of a 1- to 2-GeV proton linac as an injector into the main ring. The name of the laboratory was changed to the Lawrence Radiation Laboratory (LRL) in 1959, in honor of Ernest O. Lawrence, who died in 1958; E. M. McMillan became the new director.

At an International Conference on High-Energy Particle Accelerators held at Brookhaven in September 1961, an extensive program was arranged for the exchange of information on designs of multihundred-GeV accelerators. Scientists from CERN and the European countries attended and described their existing designs; the three U. S. laboratories also presented detailed status reports. Unfortunately, however, the USSR delegation did not arrive, so the exchange was incomplete.

The Atomic Energy Commission had been for years the primary source of supporting funds for construction of accelerator facilities and high-energy physics research in the United States. The Commission was kept informed of the planning for higher energies in the several laboratories and encouraged the exchange of ideas with scientists abroad. In November 1962 a special panel was appointed jointly by the General Advisory Committee of the AEC and the President's Scientific Advisory Committee, to study the status of high-energy physics and to recommend a program for the future. The Report [7] of this panel (called the "Ramsey panel" after its chairman, Professor N. F. Ramsey of Harvard University) was released on May 10, 1963. In addition to making general and extensive recommendations for support of the existing high-energy physics program, the Report made several specific suggestions regarding the extension of facilities into the very-high-energy range. The panel proposed a two-step approach, starting with early authorization of construction for the 200-GeV proton accelerator then being considered at the LRL, and continued support of design studies at Brookhaven of a national accelerator in the 600- to 1000-GeV range to be authorized at a later date. It also suggested development of plans for a proton-proton storage ring at Brookhaven as an intermediate step toward the study of higher-energy interactions. The recommendations of this panel justified the AEC in implementing plans at LRL and Brookhaven. Both laboratories were authorized to proceed with their design studies starting in April 1963.

THE LRL DESIGN STUDY

At the Lawrence Radiation Laboratory, authorization of the 200-GeV design study in 1963 resulted in a major effort extending over four years with an average professional staff of 35 persons. E. J. Lofgren headed the study and Lloyd Smith led the theoretical section. The culmination was an Interim Report presented to the AEC in December 1964, followed by a "Design Study for a 200-GeV Accelerator" [8] in June 1965. The design study covered the scientific, technical, and engineering features of the accelerator and the associated facilities, and included a preliminary engineering cost estimate and time schedule for completion. The cost of the facility plus basic experimental equipment was estimated to be $350 million, with a continuing annual operations cost of $50 to $60 million.

The LRL "Design Study" was a description of a single integral design that was feasible, with realistic cost estimates. During the study many alternative concepts were considered and evaluated. Although a fixed set of parameters was chosen, it was recognized that changes and improvements could be expected with further developments. To make the cost estimates meaningful, a single site was selected, in the Sierra Nevada foothills above Sacramento. It was conceived that the accelerator would be designed and its construction supervised by an expanded staff coming primarily from the LRL, but after completion it would become a National Laboratory available to all qualified scientists.

Following publication of the LRL "Design Study" and during the following year before the AEC announced the selection of another site for such an installation, work continued on possible improvements, optimization of parameters and refinement of cost estimates. A summer study held in 1966 explored further the instruments and facilities needed for experimental use. Work went on at Berkeley for two more years, until the

National Accelerator Laboratory was organized and took over the task.

Meanwhile, designs were in progress in other laboratories. At Brookhaven the design study for an accelerator in the 600- to 1000-GeV range continued, but on a considerably lower scale of effort than at LRL. Emphasis was on analysis of feasibility and general parameters for these very high energies, without much engineering detail or cost estimating. A summer study held at Brookhaven in 1963 was attended by a large number of physicists and accelerator experts from this country and abroad. This study showed that there was considerably more interest in increasing intensity of the existing AGS than in a more detailed design study for superhigh energies or for a set of proton storage rings to utilize the 30-GeV protons for beam-beam interactions. A program for conversion of the AGS to produce ten times higher intensities was initiated, and soon authorized by the AEC.

At the CERN Laboratory in Geneva, a group of design experts from the CPS division started a design study as early as 1961, with alternative initial goals of 150- and 300-GeV accelerators, and continued for the next 3 to 4 years. By the time United States planning began to focus around 200-GeV energy, the CERN planning concentrated on the 300-GeV machine. A design study was published [9] in 1964, and a proposal was submitted to the CERN Council for further planning and negotiation between the member States. A Committee was appointed by the Council to initiate studies of possible sites for a 300-GeV accelerator in western Europe.

During these years exchange of ideas continued between the design groups at LRL, Brookhaven, CERN (and interested individuals elsewhere). A series of meetings took place at approximately 6-month intervals between representatives of these groups. A few exchange visits of extended duration by group members maintained continuity in this collaborative effort. And exchange of ideas continued with Soviet scientists at international meetings, with the possible goal of an international accelerator for the superhigh-energy range.

The general conclusion from these design studies was that the basic principle of the AG proton synchrotron could be extended with certainty to the 200- or 300-GeV range or even considerably higher. Peak energy would be set by orbit radius for conventional iron-cored magnets, and would probably be determined by economic rather than by technical considerations. Another consequence of the use of very large orbits was the availability of very high beam intensities per pulse, owing to the large orbit circumference. Intensities of the order of 10^{13} protons per second could be expected, with beam power approaching 1 megawatt. The higher-intensity advantage of other techniques, such as use of fixed-field alternating gradients (FFAG) or high-duty-cycle linear accelerators, was to a great extent compensated by the large orbits, in the hundred-GeV range. Design cost estimates varied among the several laboratories, based on different engineering styles, but unit costs were in the range of $100 million to $150 million per 100 GeV energy.

SCIENTIFIC JUSTIFICATION AND GOVERNMENT POLICY

The studies at LRL, Brookhaven, and other laboratories also included surveys of the purposes of high-energy physics and theoretical justifications for higher energies. Considerable effort went into preliminary planning and feasibility studies of beam-separation and beam detection equipment in the hundred-GeV range. In December 1964, a Brookhaven Report [10] edited by L. C. L. Yuan presented statements by about 25 leading theoretical scientists in the field. These statements were unanimously favorable, and even urgent, in their advocacy of the need for new accelerators in the higher energy range.

In the spring of 1964, the National Academy of Sciences–National Research Council established a physics survey committee, under the chairmanship of George E. Pake, to study future requirements in relation to national needs in physics (and other fields of science). A subpanel on elementary particle physics, Robert Walker, chairman, brought in a report which was basically in accord with the conclusions of the Ramsey panel, but recommended that future high-energy accelerators be considered national rather than regional facilities. This report of the Pake committee was instrumental in developing national policy in the field of high-energy physics. It made basic recommendations which were repeated in a policy paper prepared within the AEC. The most significant result was the identification of future high-energy accelerators as national rather than regional facilities, with the implication that choice of site was an important aspect of this function. In subsequent actions the AEC turned down the Midwest Universities Research Association (MURA) proposal for a regional FFAG accelerator in the midwest designed to produce very high beam intensities at less than 100-GeV energy. The LRL proposal for a 200-GeV machine to be located near and operated by the University of California was reconsidered in the light of its significance as a national facility.

A study paper on "Policy for National Action in the Field of High Energy Physics," referred to above, was prepared by members of the AEC Research Division, and became the basis for a U.S. Joint Committee Print [11] published in February 1965. This policy statement reviewed progress in high-energy physics and high-energy accelerators, summarized the needs for higher energy and higher intensity, and made specific recommendations for government action. In particular, it noted the need for new high-energy facilities for the large number of user-scientists within universities. It recommended a two-step approach, starting with a 200-GeV machine similar to that designed at LRL to be located at a site convenient to a nationwide distribution of users, to be followed in the 1970's by a machine of about 1000-GeV energy.

Hearings on the high-energy research program were held before the Subcommittee on Research, Development and Radiation of the Joint Committee on Atomic Energy of the United States Congress. Many scientists from government laboratories and from universities were called as witnesses to present their views. The LRL and Brookhaven design studies were reported, and summaries entered in the testimony. High-energy physicists described the scientific justification and Yuan's collection [10] of essays by theorists was presented. AEC and other government officials presented program plans and budget estimates for the future. A full report of the hearings was published.[12]

A general conclusion from these hearings was that the scientists had made a strong case. Government officials seemed persuaded that a national facility for research in the multi-hundred-GeV range was well justified.

UNIVERSITIES RESEARCH ASSOCIATION

A group of leading scientists from the National Academy of Sciences began in early 1965 to consider the problems of management of the large scientific facility required to support research operations of a large multi-GeV accelerator. The purpose was to provide university backing and support with a broad national basis to the planning and management of the proposed 200-GeV laboratory and the subsequent research program. Frederick Seitz, President of the National Academy of Sciences, acted as coordinator of scientific opinion in calling a meeting of university presidents, first in January and again in June 1965. The outcome of these meetings was the organization of the Universities Research Association (URA), consisting originally of 34 universities distributed across the United States which carry on significant research programs in the physical sciences.

The URA is incorporated in the District of Columbia and maintains its principal office at 2100 Pennsylvania Avenue,

N.W., Washington, D.C. The Council of Presidents of the member universities meets about once a year. The president of URA is Professor Norman Ramsey of Harvard University. Active management is placed in a Board of Trustees elected by the Council of Presidents. The Trustees include a representative from each of 15 regional groups of universities and 6 at large to represent the public interest. Each member university agreed to contribute up to $100,000 to URA if and when called upon by the Trustees, primarily for the expenses of organization and operation of URA. By January 1968 the number of member universities had grown to 48.

The URA offered its services to the AEC as a management organization to contract with the AEC and to operate the 200-GeV facility. The organization was not involved in selection of the site for the laboratory, but indicated willingness to undertake construction and operation at any site selected by the federal government.

SELECTION OF THE SITE

The choice of Weston, Illinois (35 miles west of Chicago) as the site for the 200-GeV laboratory was made by the Atomic Energy Commission after extensive site-selection studies.[13]

The search for the site started in April 1965, when the AEC issued a press release inviting statements of interest in proposing sites for this new scientific facility. A total of 125 proposals were ultimately received, relating to more than 200 different locations, one or more from each of 48 states. By September 1965 the Commission had reduced the list to 85 proposals relating to 148 sites. The AEC then requested the National Academy of Sciences to enlist a site evaluation committee composed of eminent scientists to review and evaluate the site proposals and make recommendations to the AEC. The chairman of this committee was E. R. Piore. The committee visited and studied the sites which met the basic selection criteria, and in March 1966 reported to the Commission recommending six sites as of equal merit. As part of the final evaluation effort, three commissioners visited each of the six recommended sites. The Commission announced its unanimous decision for the site at Weston, Illinois, on December 16, 1966.

As was to be expected, the selection of a site was not popular with the proponents of many other sites. The choice caused considerable discussion in the Congress and in Joint Committee Hearings.[13] Much of the discussion centered on the availability of open housing in the Chicago suburban area, a problem raised by the National Committee against Discrimination in Housing and other civil rights groups. In the testimony, the state of Illinois was called upon to enact fair-housing legislation and suburban communities were importuned to the same effect. In partial answer, the Commission made clear its basic adherence to a policy of nondiscrimination.

In the hearings, the AEC proposed a reduced-scope accelerator dictated by the Bureau of the Budget for budgetary reasons. In this reduced scope the energy was to be retained at 200 GeV but the intensity could be decreased initially to about one-tenth that of the LRL design and the number of experimental target stations reduced. An initial cost figure of about $240 million was suggested, with the expectation that additional funds would be provided later to recover the original scope.

The general result of these hearings was congressional authorization to the AEC to proceed with the planning for a 200-GeV accelerator, but at the lower initial cost. The first step was the execution of a contract with the URA to initiate a design study to accomplish this purpose, which was signed on January 23, 1967.

ESTABLISHMENT OF THE NATIONAL ACCELERATOR LABORATORY

Following selection of the Weston site, the URA moved promptly to select a director and initiate activities. The first scientist asked to direct the design study declined, and Professor Robert Rathbun Wilson of Cornell University was offered the position of director of the National Accelerator Laboratory (NAL). He announced his intention to prepare a design for high intensity with a higher-energy option within one year at a cost not to exceed $240,000; but he required significant concessions of authority from the AEC and URA and a fast time schedule (5 years) for construction. These conditions were acceptable. Wilson accepted the position as director on March 7, 1967; he took up the position full time on June 15, when he moved to Chicago.

President Ramsey called a meeting of potential scientific users of the 200-GeV accelerator for April 7–8, 1967 at the Argonne National Laboratory, at which time announcements were made of the organization of NAL and the appointment of Wilson as director. Wilson used this opportunity to call a special meeting of accelerator scientists attending the meeting and announced a summer design program to start on June 15 in the Chicago area.

In April, a preliminary contract for architectural-engineering services was signed with the firm of "DUSAF," * and their engineers started meeting with the director and his associates. They were asked to perform site surveys and make preliminary site plans (Fig. 32). Also in April, arrangements were made for temporary office space in Oak Brook, Illinois.

By May a number of scientific and administrative staff members had been recruited and accepted appointments, in most cases to start at some later date. These included E. L. Goldwasser (University of Illinois) as Deputy Director, F. T. Cole (LRL), A. L. Read (Cornell), J. DeWire (Cornell), M. S. Livingston (CEA), D. E. Young (University of Wisconsin), C. D. Curtis (University of Wisconsin), and on the administrative side Donald Getz and Donald Poillon.

* DUSAF is a consortium of four firms: Daniel, Mann, Johnson, and Mendenhall; Max O. Urbahn; Seelye, Stevenson, Value, and Knecht; and George A. Fuller.

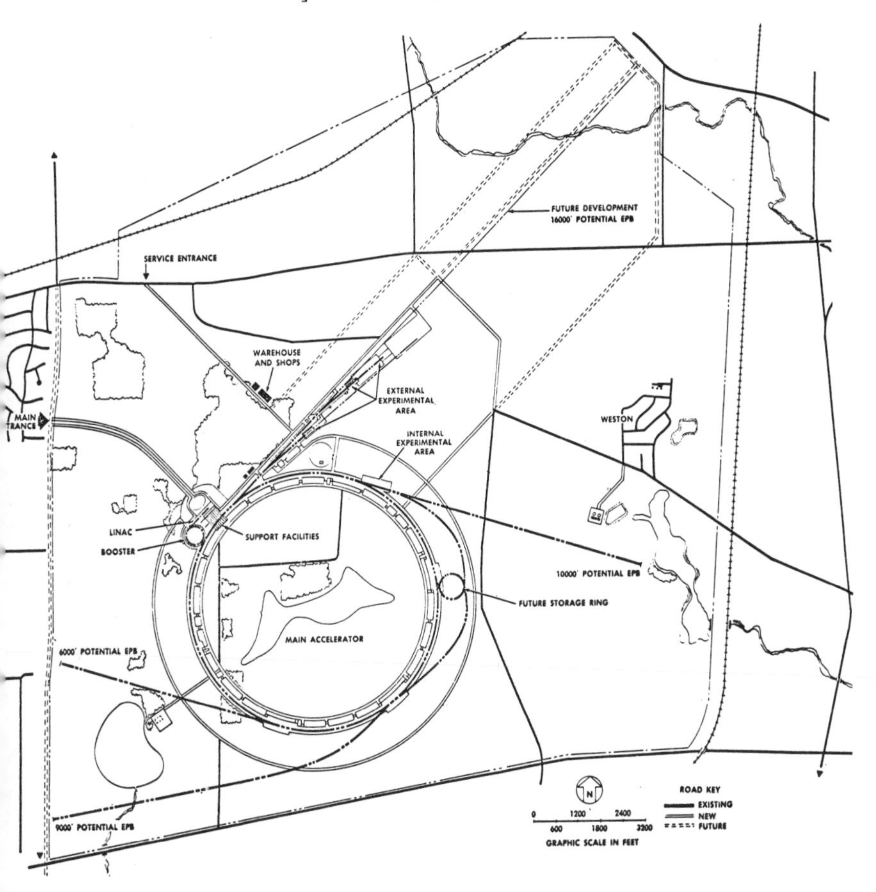

FIG. 32. Site plan for the 200–400 GeV accelerator to be built by the National Accelerator Laboratory at Weston, Illinois.

Meetings were held at an increasing rate during the spring of 1967: of the URA Trustees, of a Scientific Advisory Committee appointed by the director, with representatives of DUSAF, and with accelerator design experts from many laboratories. The AEC took a leading role in planning and in solving administrative problems, and their representatives attended all planning meetings and discussions. All obstacles were cleared away, offices for the summer program were prepared, contracts for financial support by URA and the AEC were formally signed, and commitments were obtained from accelerator scientists from many sources to attend and contribute to the summer program.

SUMMER DESIGN PROGRAM OF 1967

The purpose of the 1967 summer program was to develop concepts for a new design for a 200-GeV accelerator, significantly simpler and of lower cost than that proposed in the LRL design report. The leader in this conceptual study was the new director, R. R. Wilson, who brought to bear his recent experience in building the Cornell 10-GeV AG electron synchrotron at a lower cost per GeV than previous electron synchrotrons of this type. Some of these simpler and lower-cost concepts became part of the new design from the start, such as a compact magnet structure which embodied a girder-type support in its construction, and a minimal-size magnet tunnel enclosure without overhead crane or magnet foundation piers. Other simplifying concepts were provided by visiting accelerator experts and scientists. Several significant alternative solutions were developed and presented by visitors from LRL, which had several months' lead in searching for cost reductions following the AEC requirement of a reduced-scope accelerator discussed in the hearings in February. The theme of the study became a search for new and different solutions to design problems, with the emphasis on reduction of cost without excessive loss of quality or reliability.

The program started on June 15, 1967, with 20 accelerator scientists attending the first week, including Wilson and Goldwasser. Additional members arrived later and others came for a few weeks, with an average attendance of about 25. Credit for this development should be distributed among the large number of scientists and engineers who contributed. A total of 63 persons were involved in the study during the summer and fall, of whom 30 eventually joined the NAL staff.

The more significant features considered and ultimately accepted in the new design deserve to be identified individually:

(*a*) *Option for Higher Energy*. Well before the congressional hearings in January–February 1967, it was becoming evident that the two-step program (for a 200-GeV machine to be followed by a 600- to 1000-GeV machine) might postpone too long the attainment of truly high energies. Many scientists began to regret their commitment to 200 GeV and wished for higher energies. Also, the reduced scope specified by the AEC required rethinking about the energy limitations of a ring of magnets limited by saturation of the iron. One concept for reducing initial cost (originated by Garren, Lambertson, Lofgren, and Smith at LRL) consisted of filling half the ring with magnets, with an initial operating energy that could be increased later by the addition of other magnets; it was called the "expanditron" in laboratory slang. Prior to the summer program, this concept was modified to consider use of an orbit filled with magnets but initially powered at half field to reduce cost, with additional power supplies to be added later; this procedure would minimize the downtime required for a future conversion. Wilson adopted this concept and extended it to imply an orbit size capable of ultimate powering to reach 400 or even 500 GeV but initially powered for 200-GeV operation. He hoped that this option for higher energy could be included within the $240 million budget. The concept was discussed with enthusiasm as early as January 1967, in local groups and at URA meetings. It typified the basic theme of the new approach to design and had great appeal at AEC and congressional levels.

(*b*) *Separated-Function Magnets*. Alternating-gradient accelerator designers had long known of the option of separating the bending and focusing functions of the AG magnets in the main ring. For the large orbit (radius 1 km) required for 400 GeV the focusing properties could be provided by relatively short and widely spaced quadrupole magnets, allowing most of the orbit to be filled with bending magnets which would have uniform and very high fields, flat poles, and simplified construction. This separated-function design significantly reduced estimated costs for the magnet system.

(*c*) *Minimum Magnet Enclosure.* Experience at electron synchrotron laboratories such as CEA and Cornell which used small AG magnets has increased confidence that installation and maintenance can be accomplished in a small tunnel without an overhead crane. In typical maintenance procedures, complete magnet units (20 ft long) will be replaced by means of special handling vehicles. The desire to minimize maintenance time and reduce radiation exposure results in removing much of the auxiliary equipment from the tunnel and into separate equipment galleries. These plans justified a simply constructed and minimum-size tunnel, formed of precast concrete sections of 10-ft width (Fig. 33). The cost savings in design estimates were considerable.

(*d*) *Magnet Foundations.* With the development toward AG magnets of small cross section and light weight, confidence has increased among accelerator designers that concrete piers for the magnet are not necessary. Experience shows that magnets can be aligned with the aid of position monitors for the beams. As a consequence, magnet piers were eliminated from the design and the magnets are to be mounted directly on the slab floor of the tunnel. Although there were many critics from outside the NAL group, this concept prevailed.

(*e*) *Fast-Cycling Booster.* All early plans for a multihundred-GeV machine divided the acceleration into several steps, with one or more preaccelerator stages feeding into the main ring. A multi-GeV linear accelerator has the obvious advantage of a linear beam for injection into the ring, but has an undesirably high cost per GeV. The problem with a smaller injector synchrotron was the uncertain efficiency of ejection of the preaccelerated beam and the difficulty of injecting it into the main ring. Success in the ejection of high-quality emergent beams from the AGS and PS synchrotrons during the middle 1960's

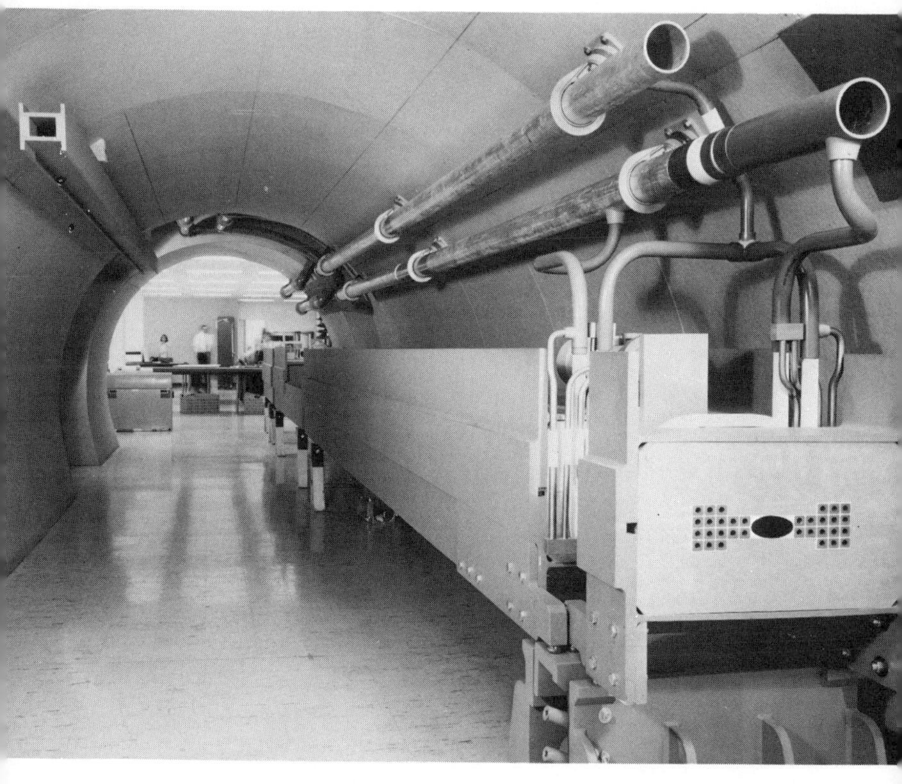

FIG. 33. Full-scale mock-up of the tunnel enclosure and a magnet sector for the 200–400 GeV accelerator.

solved this problem and made it possible to plan for the use of a 10-GeV "booster" synchrotron as an injector into the 200-GeV accelerator. The booster is operated at a fast cycling rate so that successive booster output pulses can be used to fill the main ring completely. This concept has been adopted at the NAL, with the booster operating at 15 pulses per second and a sequence of 13 pulses (0.8 sec) used to fill the main ring. The result is a high-intensity circulating beam in the main ring, at the cost of a slightly lower cyclic repetition rate (15 to 20 per minute) than might otherwise be achieved.

(*f*) *Long Straight Sections.* The need for relatively long straight sections for ejection of an emergent 200- to 400-GeV beam was obvious at the start of the design study. The accepted technique is to use matching quadrupole pairs at the ends of each straight section. Six long straights each with 160 ft of free space are planned, to be used for injection, ejection, and future options.

(*g*) *Vertical Injection and Ejection.* The compact magnet design, with small vertical height above and below the beam has offered an opportunity of injecting the beam from the linac into the booster in the vertical rather than the horizontal plane. The same principle is used for ejection from the booster and injection into the main ring. Orbit analysis shows that such vertical injection gives somewhat larger beam acceptance in the main ring. It also removes injection equipment from the radial plane where it might be damaged by radiation or induced radioactivity produced by the main beam.

(*h*) *Single Emergent Proton Beam.* A number of members of the summer program were experimental high-energy physicists who devoted their efforts to the planning of experiments and the supporting facilities. Originally, several emergent-beam facilities had been planned, fanning out from successive long straight sections. The concept of using a single emergent beam, with switching magnets to utilize three or more target stations, was originally introduced as a means of reducing initial costs. However, as the study progressed, excellent reasons were developed to justify such a single emergent beam run as the total external beam facility. With one extended beam path, communications and transportation of equipment become simpler, and ejection efficiency can be made higher, than with several shorter paths. The location of the main ring and emergent-beam run on the site were arranged to maximize accessibility to the experimental facilities.

(*i*) *Options for the Future.* The major future option is that for increasing power to operate at 400- or even 500-GeV energy.

It was also considered essential to include the option of adding a colliding-beam facility in the future. The main-ring location and the single emergent-beam run allowed plans for a future storage-ring facility to be fed from other, unused long straight sections. Options include the use of bypass sectors external to the main ring and storage rings of either small or large diameter.

By September 15 most of the basic concepts of the new design had been crystallized, at least in principle. Preliminary descriptions and cost estimates were prepared and circulated to critics in other laboratories; several discussion sessions were held at which criticisms were considered and changes made when they proved valid. A number of critics remained unsatisfied, but no alternative concepts were proposed that would not significantly increase the estimated costs. As director, Wilson made a sequence of basic decisions freezing one by one the major concepts and parameters. Visitors from other laboratories returned to their home bases; NAL staff members were enlisted and joined the continuing staff. The total staff at design headquarters during the fall did not drop below the 25 averaged during the summer program.

Activities initiated during the summer continued through the fall, with increasing emphasis on more detailed designs, improved parameters, and improved cost estimates. The laboratory organization took form and business and clerical staff were added. A machine shop was initiated and installed in quarters in Downers Grove, 5 miles from Oak Brook, and additional office space was leased. The professional staff grew to 30 by December 30, with 8 appointees to start after the first of the year. Total laboratory staff, including offers outstanding, was 90 on December 30.

One initial deadline for the laboratory was preparation of the AEC Construction Data Sheet (Schedule 44), to be submitted by October 15, 1967. This represented the final and complete cost estimate for construction. It was prepared and submitted on time, with a total construction-cost estimate of $242 million. As a result of schedule revisions required by budgetary limitations, the AEC changed and rounded off the estimated cost to $250 million.

The final product of the design program was a design report to the AEC presenting the results of the study, detailed parameters and justifications, and a cost estimate. The original date requested by the AEC Research Division for this report was December 15. On that date a final copy of the "200 BeV Accelerator, Parameters and Specifications" [14] was completed; additional copies were available for general distribution before January 15, 1968. This design report was submitted to the Joint Committee on Atomic Energy and was incorporated in the AEC Authorizing Legislation for fiscal 1969, published in a U.S. Government Print [15] in February 1968. The bill authorizing start of construction of the 200-GeV accelerator was passed by the Congress and was signed into law by the President on April 19, 1968.

[Appendix]

Chronology of Development of Particle Accelerators

This listing gives names of investigators or groups, dates of new concepts and of first operations of new types, and new energy records.

PREHISTORY

1919–1921 Schenkel, Greinacher; voltage-multiplier circuit.

1924 Ising; proposal of electron linac with drift tubes, spark-gap excitation.

1925 Sorensen; cascade transformer at 750 kV for ac electrical testing.

1928 Lauritsen; cascade transformer for X-rays at 750 kV.
Wideröe; radiofrequency resonance with heavy ions in two-step linac.

1930 Brasch and Lange; surge generator, 2.4 MV, discharge in chamber.
Tuve *et al.;* Tesla-coil resonance transformer, 1–2 MV.
Van de Graaff; electrostatic generator, 1.5 MV.

1931 Lawrence and Livingston; demonstration of cyclotron resonance.
Sloan and Lawrence; linear accelerator, 1.2-MeV Hg ions.

1932 Bellaschi; surge generator, 6 MV, for ac testing.
Urban *et al.;* thunderstorm potentials in Alps.

EARLY HISTORY

1932 Cockcroft and Walton; voltage multiplier, 0.5-MeV protons; disintegration.
Lawrence and Livingston; cyclotron, 1.2-MeV protons; disintegration.

1933 Tuve, Hafstad, and Dahl; electrostatic generator, 0.6-MeV protons.

1934 Lawrence and Livingston; cyclotron, 5.0-MeV deuterons.
Tuve, Hafstad, and Dahl; electrostatic generator, 1.2-MeV protons and deuterons.

1935 Herb *et al.;* pressure electrostatic generator, 0.7-MeV protons.

1936 Lawrence and Cooksey; cyclotron, 8.0-MeV deuterons.

1938 Herb *et al.;* pressure electrostatic generator, 2.7-MeV protons and deuterons.
Bethe and Rose; relativistic limitation of cyclotron.
Thomas; proposal of sector focusing for cyclotron.
1939 Lawrence *et al.;* 60-in. cyclotron, 16-MeV deuterons, 32-MeV He ions (later 20-MeV deuterons, 40-MeV He ions).
1940 Kerst; betatron, 2.3-MeV electrons.
1942 Kerst *et al.;* betatron, 20-MeV electrons.

RECENT HISTORY (HIGH-ENERGY ACCELERATORS)

1945 Veksler; synchronous-accelerator proposal.
McMillan; synchronous-accelerator proposal.
1946 Goward and Barnes; electron synchrotron, 8-MeV electrons.
Lawrence *et al.;* 184-in. synchrocyclotron, 200-MeV deuterons, 400-MeV He ions.
1947 Lawson *et al.;* electron synchrotron, 70-MeV electrons.
Hansen *et al.;* Stanford Mark I linac, 6-MeV electrons.
1949 McMillan; electron synchrotron, 320-MeV electrons.
1950 Stanford Mark II linac, 35-MeV electrons.
Kerst; betatron, 300-MeV electrons.
Alvarez *et al.;* linear accelerator, 32-MeV protons.
1952 Brookhaven National Laboratory; proton synchrotron (cosmotron), 2.2-GeV (later 3.0-GeV) protons.
Courant, Livingston, and Snyder; proposal of alternating-gradient focusing.
1953 Wilson *et al.;* Cornell AG electron synchrotron, 1.2-GeV electrons.
1954 Lawrence Radiation Laboratory; proton synchrotron (bevatron), 6.2-GeV protons.
1957 Lawrence Radiation Laboratory; synchrocyclotron (rebuilt), 720-MeV protons.
Dubna, USSR; synchrophasotron (proton synchrotron), 10-GeV protons.
1959 CERN; AG proton synchrotron, CPS, 28-GeV protons.
1960 Brookhaven National Laboratory; AG proton synchrotron AGS, 33-GeV protons.
Stanford Mark IV linac, 1.0-GeV electrons.
1961 High Voltage Engineering Co.; tandem Van de Graaff generator, 12 MeV.
1962 Argonne National Laboratory; ZGS proton synchrotron, 12.5 GeV.

Livingston *et al.;* Cambridge AG electron synchrotron, 6-GeV electrons.

1966 Stanford 2-mile electron linear accelerator, SLAC, 20 GeV.

1967 Cornell; AG electron synchrotron, 10-GeV electrons.

Serpukov, USSR: AG proton synchrotron, 76-GeV protons.

National Accelerator Laboratory; start of 200–400-GeV accelerator design.

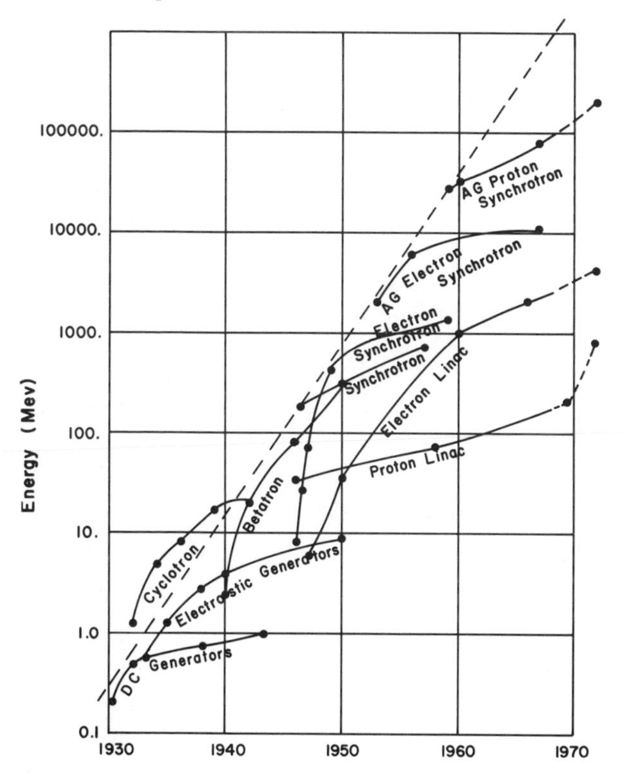

FIG. 34. Energies achieved by accelerators from 1932 to 1968. The linear envelope of the individual curves shows an average tenfold increase in energy every 6 years.

References

CHAPTER 1. THE RACE FOR HIGH VOLTAGE

1. E. Rutherford, *Proc. Roy. Soc. (London) A117,* 300 (1927).
2. P. L. Bellaschi, *Trans. Am. Inst. Elec. Engrs. 51,* 936 (1932).
3. A. Brasch and F. Lange, *Naturwiss. 18,* 769 (1930); *Z. Physik 70,* 10 (1931).
4. R. W. Sorensen, *J. Am. Inst. Elec. Engrs. 44,* 373 (1925).
5. R. Crane and C. C. Lauritsen, *Rev. Sci. Instr. 4,* 118 (1933).
6. M. A. Tuve, L. R. Hafstad, and O. Dahl, *Phys. Rev. 35,* 1406; *36,* 1261 (1930).
7. D. H. Sloan, *Phys. Rev. 47,* 62 (1935).
8. E. E. Charlton, W. F. Westendorp, L. E. Dempster, and G. Hotaling, *J. Appl. Phys. 10,* 374 (1934).
9. H. Greinacher, *Z. Physik 4,* 195 (1921).
10. G. Gamow, *Z. Physik 52,* 510 (1929).
11. E. U. Condon and R. W. Gurney, *Phys. Rev. 33,* 127 (1929).
12. J. D. Cockcroft and E. T. S. Walton, *Proc. Roy. Soc. (London) A136,* 619 (1932); *A137,* 229 (1932); *A144,* 704 (1934).
13. R. J. Van de Graaff, *Phys. Rev. 38,* 1919A (1931).
14. R. J. Van de Graaff, K. T. Compton, and L. C. Van Atta, *Phys. Rev. 43,* 149 (1933).
15. L. C. Van Atta, D. L. Northrup, R. J. Van de Graaff, and C. M. Van Atta, *Rev. Sci. Instr. 12,* 534 (1941).
16. M. A. Tuve, L. R. Hafstad and O. Dahl, *Phys. Rev. 48,* 315 (1935).
17. L. R. Hafstad and M. A. Tuve, *Phys. Rev. 47,* 506 (1935); *48,* 306 (1935).
18. R. G. Herb, D. B. Parkinson, and D. W. Kerst, *Rev. Sci. Instr. 6,* 261 (1935).
19. D. B. Parkinson, R. G. Herb, E. J. Bernet, and J. L. McKibben, *Phys. Rev. 53,* 642 (1938).
20. R. G. Herb, C. M. Turner, C. M. Hudson, and R. E. Warren, *Phys. Rev. 58,* 579 (1940).
21. J. G. Trump and R. J. Van de Graaff, *Phys. Rev. 55,* 1160 (1939).
22. R. Wideröe, *Arch. Elektrotech. 21,* 387 (1928).

23. E. O. Lawrence and N. E. Edlefsen, *Science 72,* 376 (1930).

24. M. S. Livingston, "The production of high-velocity hydrogen ions without the use of high voltages," Ph.D. thesis, University of California, April 14, 1931.

25. E. O. Lawrence and M. S. Livingston, *Phys. Rev. 40,* 19 (1932).

26. E. O. Lawrence, M. S. Livingston, and M. G. White, *Phys. Rev. 42,* 1950 (1932).

27. E. O. Lawrence and M. S. Lvingston, *Phys. Rev. 45,* 608 (1934).

28. E. O. Lawrence, L. W. Alvarez, W. M. Brobeck, D. Cooksey, D. R. Corson, E. M. McMillan, W. W. Salisbury, and R. L. Thornton, *Phys. Rev. 56,* 124 (1939).

29. D. W. Kerst, *Phys. Rev. 58,* 841 (1940).

30. D. W. Kerst, G. D. Adams, H. W. Koch, and C. S. Robinson, *Rev. Sci. Instr. 21,* 462 (1950).

CHAPTER 2. ERNEST LAWRENCE AND THE CYCLOTRON

1. E. O. Lawrence and N. E. Edlefsen, "On the production of high speed protons," *Science 72,* 376 (1930).

2. R. Wideröe, *Arch. Elektrotech. 21,* 387 (1928).

3. E. O. Lawrence and M. S. Livingston, *Phys. Rev.* (A) *37,* 1707 (1931).

4. M. S. Livingston, "The production of high-velocity hydrogen ions without the use of high voltages," Ph.D. thesis, University of California, April 14, 1931.

5. E. O. Lawrence and M. S. Livingston, *Phys. Rev.* (L) *38,* 834 (1931); E. O. Lawrence and M. S. Livingston, *Phys. Rev.* (A) *38,* 862 (1931); E. O. Lawrence and M. S. Livingston, *Phys. Rev. 40,* 19 (1932).

6. J. D. Cockcroft and E. T. S. Walton, *Proc. Roy. Soc.* (*London*) *A137,* 229 (1932).

7. E. O. Lawrence, M. S. Livingston and M. G. White, *Phys. Rev.* (L) *42,* 150 (1932).

8. M. S. Livingston, *Phys. Rev.* (L) *42,* 441 (1932); M. S. Livingston and E. O. Lawrence, *Phys. Rev.* (A) *43,* 212 (1933); M. S. Livingston, *Phys. Rev.* (A) *43,* 214 (1933); E. O. Lawrence and M. S. Livingston, *Phys. Rev. 45,* 608 (1934).

9. M. S. Livingston, M. C. Henderson and E. O. Lawrence, "Neutrons from deutons and the mass of the neutron," *Phys. Rev.* (L) *44,* 782 (1933).

10. M. C. Henderson, M. S. Livingston and E. O. Lawrence, "Artificial radioactivity produced by deuton bombardment," *Phys. Rev.* (L) *45*, 497 (1934).

11. E. O. Lawrence and D. Cooksey, *Phys. Rev. 50*, 1131 (1936).

12. E. O. Lawrence, L. W. Alvarez, W. M. Brobeck, D. Cooksey, D. R. Corson, E. M. McMillan, W. W. Salisbury and R. L. Thornton, *Phys. Rev. 56*, 124 (1939).

CHAPTER 3. SYNCHRONOUS ACCELERATORS AND HOW THEY GREW

1. V. Veksler, *J. Phys. (U.S.S.R.) 9*, 153 (1945).

2. E. M. McMillan, *Phys. Rev. 68*, 143 (1945).

3. E. M. McMillan, *Phys. Rev. 68*, 144 (1945).

4. F. K. Goward and D. E. Barnes, *Nature (London) 158*, 413 (1946).

5. F. R. Elder, A. M. Gurewitsch, R. V. Langmuir, and H. C. Pollock, *J. Appl. Phys. 18*, 810 (1947).

6. D. Bohm and L. Foldy, *Phys. Rev. 70*, 249 (1946).

7. H. A. Bethe and M. E. Rose, *Phys. Rev. 52*, 1254 (1937).

8. L. H. Thomas, *Phys. Rev. 54*, 580, 588 (1938).

9. J. R. Richardson, K. R. MacKenzie, E. J. Lofgren, and B. T. Wright, *Phys. Rev. 69*, 699 (1946).

10. W. M. Brobeck, E. O. Lawrence, K. R. MacKenzie, E. M. McMillan, R. Serber, D. C. Sewell, K. M. Simpson, and R. L. Thornton, *Phys. Rev. 71*, 449 (1947).

11. M. L. Oliphant, J. S. Gooden and G. S. Hyde, *Proc. Roy. Soc. (London) 59*, 666 (1947); J. S. Gooden, H. H. Jensen, and J. L. Symonds, *Proc. Roy. Soc. (London) 59*, 677 (1947).

12. W. M. Brobeck, *Rev. Sci. Instr. 19*, 545 (1948).

13. M. S. Livingston, *Phys. Rev. 73*, 1258 (1948).

14. M. S. Livingston, J. P. Blewett, G. K. Green, and L. J. Haworth, *Rev. Sci. Instr. 21*, 7 (1950).

15. E. J. Lofgren, *Science 111*, 295 (1950).

16. G. Ising, *Arkiv Mat. Astron. Fysik. 18*, 45 (1925).

17. J. W. Beams and L. B. Snoddy, *Phys. Rev. 44*, 784 (1933), J. W. Beams and H. Trotter, Jr., *Phys. Rev. 45*, 849 (1934).

18. D. H. Sloan and E. O. Lawrence, *Phys. Rev. 38*, 2021 (1931).

19. D. H. Sloan and W. M. Coates, *Phys. Rev. 46*, 539 (1934).

20. E. L. Ginzton, W. W. Hansen and W. R. Kennedy, *Rev. Sci. Instr. 19*, 89 (1948).

21. Linear Accelerator Issue, *Rev. Sci. Instr. 26* (February 1955).

22. V. Veksler, *Compt. Rend. Acad. Sci. U.S.S.R. 43,* 444 (1944); *44,* 392 (1944); *J. Phys.* (*U.S.S.R.*), *9,* 153 (1945).

CHAPTER 4. THE STORY OF ALTERNATING-GRADIENT FOCUSING

1. L. H. Thomas, *Phys. Rev. 54,* 580, 588 (1938).
2. E. D. Courant, M. S. Livingston, and H. S. Snyder, *Phys. Rev. 88,* 1190 (1952).
3. J. P. Blewett, *Phys. Rev. 88,* 1197 (1952).
4. J. B. Adams, M. G. N. Hine, and J. D. Lawson, *Nature* (*London*) *171,* 926 (1953).
5. N. C. Christofilos, "Focusing systems for ions and electrons and applications in magnetic resonance particle accelerators" (1950, unpublished).
6. E. D. Courant, M. S. Livingston, H. S. Snyder, and J. P. Blewett, *Phys. Rev. 91,* 202 (1953).
7. "Design study for a 15-BeV accelerator," *M.I.T. Laboratory for Nuclear Science, Report No. 60* (June 30, 1953).

CHAPTER 6. THE 200-GeV ACCELERATOR

* Taken largely from M. S. Livingston, "Early History of the 200-GeV Accelerator," National Accelerator Laboratory Report, NAL-12-0100 (June 18, 1968).

1. David L. Judd, "Development from 1952–60 at LRL of planning for a larger accelerator," UCRL Memo (September 29, 1960).
2. M. Sands, "A proton synchrotron for 300 GeV," MURA Report 465 (August 1959).
3. R. R. Wilson, *Science 133,* 1602 (1961).
4. Brookhaven National Laboratory Report, BNL 772 (T-290) (August 28, 1961).
5. California Institute of Technology—Design Study Reports: CTSL-10, Matthew Sands, "A proton synchrotron for 300 GeV" (September 1960); CTSL-11, Robert I. Hulsizer, "Magnet positioning problems" (October 20, 1960); CTSL-12, M. H. Blewett, "Preliminary cost estimate" (October 20, 1960); CTSL-14, Matthew Sands, "Injection criteria" (January 5, 1961); CTSL-15, Kenneth W. Robinson, "A radiofrequency system" (January 19, 1961); CTSL-16, R. L. Walker, "Beam transfer in the cascade synchrotron" (January 1961); CTSL-21, A. V. Tollestrup, "RF synchronization during beam transfer" (March 21, 1961); CTSL-24, Vincent Z. Peterson,

"Vacuum requirements" (April 11, 1961); CTSL-25, Jon Mathews, "Effects of magnet non-linearities on betatron frequencies" (April 3, 1961).

6. D. L. Judd and Lloyd Smith, "Summary of accelerator studies since June 1960 at LRL," UCRL Memo (December 26, 1961).

7. Reference 12, Appendix 1.

8. "Design study for a 200-BeV accelerator," vols. I and II, UCRL-16,000 (June 1965).

9. "Design study of a 300-GeV proton synchrotron," vols. I and II, CERN AR/Int. SG-15 (November 19, 1964).

10. "Nature of matter—Purposes of high energy physics," ed. L. C. L. Yuan, Brookhaven National Laboratory Report, BNL 888 (T-360) (January 1965).

11. Joint Committee on Atomic Energy, Print 42-613 (Government Printing Office, Washington, D.C., February 1965).

12. Hearings—Joint Committee on Atomic Energy, Print 46-982 (Government Printing Office, Washington, D.C., March 2–5, 1965).

13. Hearings—Joint Committee on Atomic Energy, Print 76-329 (Government Printing Office, Washington, D.C., January-February 1967).

14. "200-BeV accelerator, parameters and specifications," National Accelerator Laboratory Design Report (January 1968).

15. Hearings—Joint Committee on Atomic Energy, Part 2, Print 90-281-0 (Government Printing Office, Washington, D.C., February 7 and 21, 1968).

Index